COMPLETE GUIDE TO ELECTRONIC TEST EQUIPMENT AND TROUBLESHOOTING TECHNIQUES

COMPLETE GUIDE TO ELECTRONIC TEST EQUIPMENT AND TROUBLESHOOTING TECHNIQUES

John Douglas-Young

Parker Publishing Company, Inc., West Nyack, New York

Library of Congress Cataloging in Publication Data

Douglas-Young, John.
Complete guide to electronic test equipment and troubleshooting techniques.

Includes index.
1. Electronic instruments. 2. Electronic apparatus and appliances--Testing. 3. Electronic apparatus and appliances--Maintenance and repair. I. Title.
TK7878.D58 621.3815'48 75-16173
ISBN 0-13-160085-0

Printed in the United States of America

HOW THIS BOOK CAN BE HELPFUL TO YOU

This book is a comprehensive, *practical* guide to modern electronic test equipment and how to use it when troubleshooting. It includes detailed guidelines on the characteristics, application, operation and general circuitry of instruments that represent the most helpful test equipment in use today.

Obviously, merely to provide descriptive data on test instruments in all catalogs would require a multi-volume encyclopedia. Fortunately, there is another and, for our purposes, a better way. This book approaches measurement from the direction of *what* is being measured. Since the areas of measurement—voltage, current, resistance, frequency and so on—are not as numerous and do not change, test equipment can be categorized according to its use. There is a family resemblance between individual instruments in various groups, so that only one or two in each group need to be studied in detail.

This book will be an invaluable supplement to the manufacturer's operating manual for individual electronic instruments (which the user should study before even switching on). We will cover all the most important types of test equipment, and provide exact steps to follow in their use and operation. Chapters 1 and 2 offer simplified guidelines on the key principles of measurement and measurement equipment that are basic to *all* test instruments. Following chapters will cover equipment and specific ways to test components such as resistors, coils, capacitors, vacuum tubes and semiconductors in troubleshooting. Also included is practical information on electronic instruments used for performance-testing everything from the human brain to the automobile engine. Chapters 12 and 13 are devoted specially to radio, television and hi-fi testing equipment and their use. These are followed by a chapter on laboratory calibration standards and a special section on tested shortcuts in troubleshooting, or to use the current jargon, "malfunction isolation."

This book will be particularly helpful to technicians, engineers, advanced hobbyists and experimenters. It will also be invaluable to servicing personnel who maintain test equipment and service a variety of electronic products. Whether the technician works in a laboratory or service shop, his concerted efforts help to insure that the engineer does not labor in vain. This book is cordially dedicated to all those who recognize the challenge and extraordinary opportunities offered by the use, maintenance and *accurate* measurement of electronic devices.

John Douglas-Young

CONTENTS

1

DECIDING WHAT TO MEASURE–AND HOW TO INTERPRET THE RESULTS

Electronic test instruments are devices that enable us to examine physical events which are not apparent to our senses. They convert one kind of stimulus, to which we are normally insensitive, into another that we can detect with our eyes or ears. For example, a neon tester converts invisible current into light.

A neon tester tells us if an electrical circuit is "live," which is a good thing to know before starting work on it. But it does no more than give a simple visible signal. It does not convert the signal into a numerical value.

Yet we cannot answer the simplest technical question about anything without facts and figures. "When you can measure what you are speaking about and express it in numbers, you know something about it," said Lord Kelvin. Without measurements we can't say we know what we are speaking about.

Electronic test equipment measures the behavior of free electrons. Their behavior is determined by the nature of the circuit or component part in which they are present. Under normal conditions they act in a certain way; under abnormal conditions they act differently. By suitable measurements we can find out exactly what's happening.

TEST EQUIPMENT ACCURACY

To measure electron performance, all test equipment must use a portion of the electron energy in the circuit. If this is abundant, a substantial sample can be taken with negligible effect. But where the energy is small, as is usually the case, only a small sample can be taken without disturbing the operation of the circuit, and so obtaining a false reading.

This is called *loading* the circuit. In Figure 1.1 you can see what happens when a voltmeter with a DC resistance of 10 kilohms is used to measure the voltage across a 10-kilohm resistor. Before connecting the meter, the voltage between points A and B was 40V (4mA x 10kΩ). However, when the meter is connected, the resistance between A and B will be only 5 kilohms. This allows the current in the circuit to increase to 5 milliamperes, and results in the potential drop between A and B changing to 25V (5mA x 5kΩ), which is a serious error.

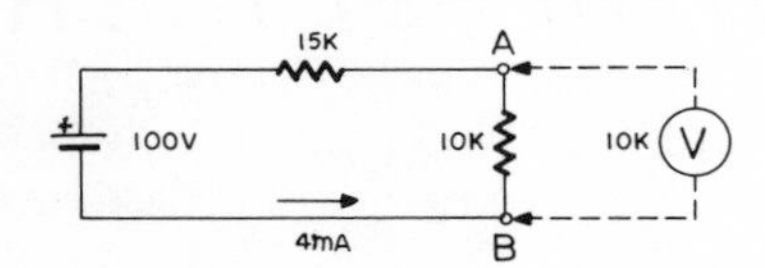

Figure 1.1 Loading Effect of a Meter

Obviously, we should have used a meter with less loading effect. One with a DC resistance of 10 megohms would cause negligible circuit disturbance and give us an accurate measurement, because it would change the potential drop between A and B by only 0.04 volt. The less sensitive meter would be fine for measuring power-line voltages (it's not going to bother the power station any!), but in a circuit where the electron flow is small, a more sensitive instrument should be used.

This is even more noticeable when measuring AC voltages, because the AC resistance of a meter decreases as the frequency increases. To obtain accurate results you would have to make sure your meter had a high *input impedance* at the frequency you are measuring. You can see that the choice of test instrument has a lot to do with the accuracy of the measurement: you can't weigh potatoes and diamonds with the same pair of scales!

The loading effect of an instrument is not the only thing affecting the accuracy of its measurements. For instance, friction in the movement of a meter may cause the pointer to stop at a different place on the dial, even though measuring the same quantity each time. This is an example of *random error*. It is not limited to "sticky" meters, but occurs to some extent in all test equipment. Since *precision* of measurement means repeatability of readings, we can only obtain it by reducing random error as much as possible.

On the other hand, imbalance in the meter movement might result in a constant offset of all readings, so that we would have precision (repeatability) but not accuracy. This kind of inaccuracy is called *systematic error*. Careful design and calibration are required to narrow the gap between measured values and true values caused by random and systematic errors.

But even when all errors have been reduced to a minimum, there will still be left some difference between real and indicated values. This difference is generally expressed as a percentage. For example, a meter may be said to have an accuracy of "plus or minus (±) 2 percent of *full scale*." This means that its readings will be always within 2 percent of the maximum value that can be indicated on the scale. If the full-scale value is 100 volts, then no reading anywhere on the scale will vary by more than 2 volts from the true value.

The accuracy of some instruments is expressed as a plus-or-minus percentage of *reading*. For instance, a signal generator might be said to have an accuracy of 5 percent of reading, which would mean that the output frequency would always be within 5 percent of the dial indication.

You will find later that the accuracy of certain types of instrument, because of their nature or use, may have to be stated in different terms, but this does not alter the point we are discussing: namely, that all instruments have some error, and their error is usually expressed as a percentage.

STANDARDS AND UNITS

But if all the clocks are off, how do you know the right time? If all readings differ somewhat from the actual values, how does anyone know the true value of anything?

The responsibility for establishing basic standard values to maintain reliability and uniformity in measurements rests with the International Bureau of Weights and Measures. More than 40 nations have so far become signatories of the treaty establishing this organization. We are represented on its committee and at its periodic conventions by the Director of the U. S. National Bureau of Standards.

The International Bureau has defined measurement units used in electronics in terms of the basic units of length, mass and time. The definitions of the basic units are:

Length: the *meter* (m) is the length equal to 1 650 763.73 wavelengths in vacuum of the radiation corresponding to the transition between the levels $2p_{10}$ and $5d_5$ of the krypton-86 atom.

Mass: the *kilogram* (kg) is equal to the mass of the international prototype of the kilogram. This is a cylinder of platinum-iridium alloy preserved in a vault at Sèvres, France, by the International Bureau of Weights and Measures.

Time: the *second* (s) is the duration of 9 192 631 770 periods of radiation corresponding to the transition between the two hyperfine levels of the ground state of the cesium-133 atom.

These "prototype units," as you can see, are based upon physical quantities that never change, as far as we know. The unit of force, the *newton* (N), is a "derived unit," because it is derived mathematically from the prototype units, being that force which gives to a mass of 1 kilogram an acceleration of 1 meter per second per second.

The *ampere* (A) is also a derived unit, and is defined as that constant current which, if maintained in two straight conductors of infinite length and of negligible cross section, and placed 1 meter apart in vacuum, would produce between these conductors a force equal to 2×10^{-7} newton per meter of length.

Actually, you couldn't really get "two straight parallel conductors of infinite length, [and] negligible cross section," but the definition shows how the ampere is derived mathematically from the basic units of length, mass and time. The other electrical units are also derived units, related to the prototype units and to each other as follows:

TABLE I—ELECTRICAL UNITS

Unit	*Quantity*	*Derivation*
Coulomb (C)	Charge	amperes × seconds (A.s)
Farad (F)	Capacitance	amperes × seconds, divided by volts (A.s/V)
Henry (H)	Inductance	volts x seconds, divided by amperes (V.s/A)
Hertz (Hz)	Frequency	reciprocal of seconds (s^{-1})
Joule (J)	Work	newtons × meters (N.m)
Ohm (Ω)	Resistance	volts divided by amperes (V/A)
Volt (V)	Voltage	watts divided by amperes (W/A)
Watt (W)	Power	joules per second (J/s)

Some of these units have replaced older ones. For instance, hertz have taken the place of cycles per second (cps) and newtons have displaced dynes. In applications where these units are inconveniently large or small, prefixes are joined to them to form multiples or submultiples, as shown in Table II.

TABLE II—STANDARD PREFIXES

Prefix	*Meaning*	*Scientific Notation*
tera- (T)	one million million	1×10^{12}
giga- (G)	one thousand million	1×10^{9}
mega- (M)	one million	1×10^{6}
kilo- (k)	one thousand	1×10^{3}
hecto- (h)	one hundred	1×10^{2}
deca- (da)	ten	1×10^{1}
deci- (d)	one-tenth	1×10^{-1}
centi- (c)	one-hundredth	1×10^{-2}
milli- (m)	one-thousandth	1×10^{-3}
nano- (n)	one-thousandth of a millionth	1×10^{-9}
pico- (p)	one-millionth of a millionth	1×10^{-12}
femto- (f)	one-thousand millionth of a millionth	1×10^{-15}
atto- (a)	one-million millionth of a millionth	1×10^{-18}

When one of these units is attached to the basic unit, it is the same as if the basic unit were being used with the corresponding power of ten. For instance:

30 kilometers (30 km) = 30×10^{3} meters (30,000 meters);
1 milliampere (1 mA) = 1×10^{-3} ampere (0.001 ampere);
5 picofarads (5 pF) = 5×10^{-12} farad (0.000000000005 farad).

Further examples are given in Table III, which lists frequencies and wavelengths in the electromagnetic spectrum. The first column gives cycles per second in power of ten (10^{n}). The second column gives the frequencies in hertz with those prefixes used in practice (hecto-, deca-, deci- and centi are not used with electronic units.) The third, fourth and fifth columns give the corresponding wavelengths in meters, microns and angstroms. The right-hand column names the type of radiation and gives some examples of its use.

Some people have difficulty in deciding whether to write "terohm" or "teraohm," although they always write "megohm." The rule is that you drop the *a* from prefixes ending in *a* before any unit beginning with a vowel. Prefixes ending in *o* drop the *o* before units beginning with *o*, but prefixes ending with *i* do not drop it before any vowel.

Because people prefer to handle smaller numbers whenever possible, you will find values in the lower part of the spectrum are usually stated in terms of frequency, those in the upper part in terms of wavelength.

You will also be coming across the word meter with the spelling metre, as it has now been decided that the United States should conform to the spelling used by other countries. However, it will take some time for it to percolate everywhere.

Some of the abbreviations (not the full names) of the metric units have capital letters. These are units derived from proper names, such as Hz (hertz) from Hertz and A (ampere) from Ampère.

TABLE III—ELECTROMAGNETIC SPECTRUM

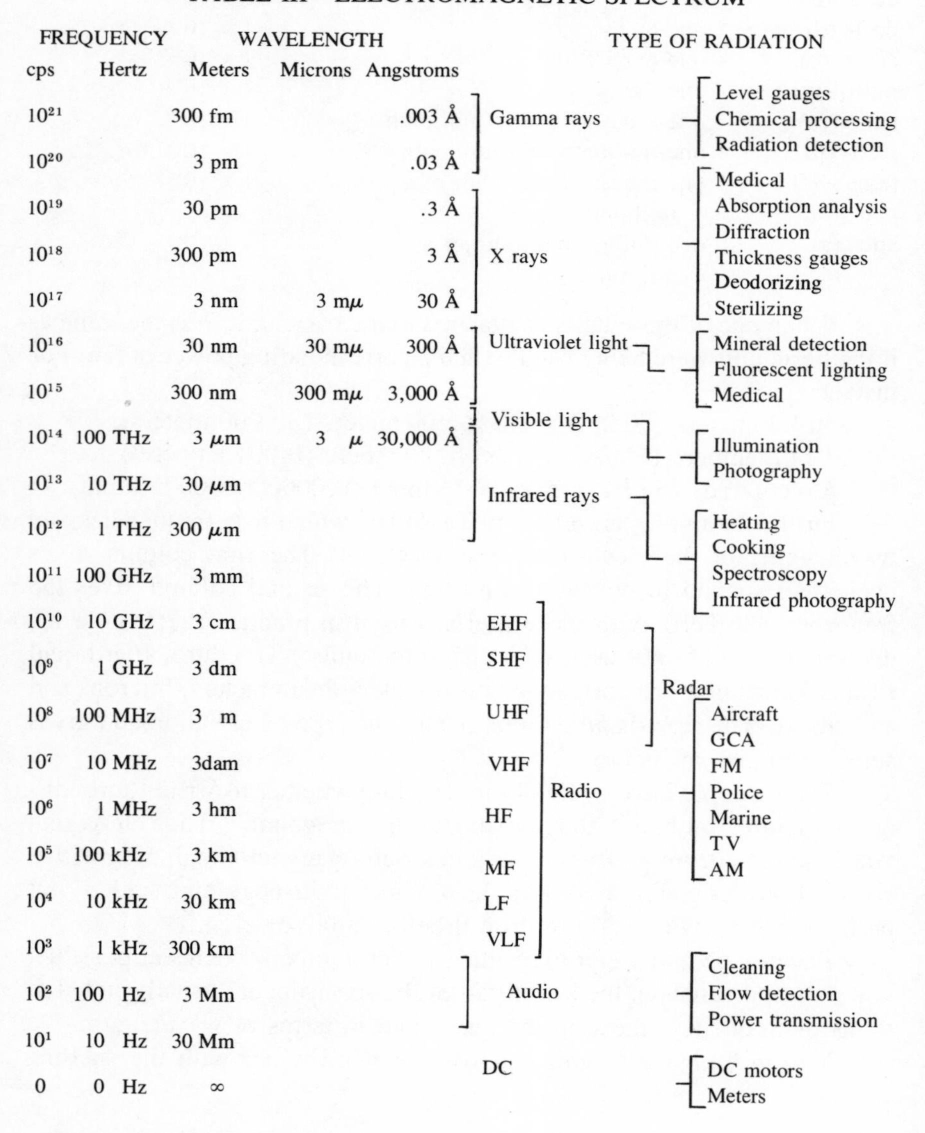

FREQUENCY		WAVELENGTH			TYPE OF RADIATION		
cps	Hertz	Meters	Microns	Angstroms			
10^{21}		300 fm		.003 Å	Gamma rays		Level gauges, Chemical processing, Radiation detection
10^{20}		3 pm		.03 Å			
10^{19}		30 pm		.3 Å	X rays		Medical, Absorption analysis, Diffraction, Thickness gauges, Deodorizing, Sterilizing
10^{18}		300 pm		3 Å			
10^{17}		3 nm	3 mμ	30 Å			
10^{16}		30 nm	30 mμ	300 Å	Ultraviolet light		Mineral detection, Fluorescent lighting, Medical
10^{15}		300 nm	300 mμ	3,000 Å			
10^{14}	100 THz	3 μm	3 μ	30,000 Å	Visible light		Illumination, Photography
10^{13}	10 THz	30 μm			Infrared rays		Heating, Cooking, Spectroscopy, Infrared photography
10^{12}	1 THz	300 μm					
10^{11}	100 GHz	3 mm					
10^{10}	10 GHz	3 cm			EHF	Radio	Radar
10^{9}	1 GHz	3 dm			SHF		
10^{8}	100 MHz	3 m			UHF		Aircraft, GCA, FM, Police, Marine, TV, AM
10^{7}	10 MHz	3dam			VHF		
10^{6}	1 MHz	3 hm			HF		
10^{5}	100 kHz	3 km			MF		
10^{4}	10 kHz	30 km			LF		
10^{3}	1 kHz	300 km			VLF		
10^{2}	100 Hz	3 Mm			Audio		Cleaning, Flow detection, Power transmission
10^{1}	10 Hz	30 Mm					
0	0 Hz	∞			DC		DC motors, Meters

MEASUREMENT LEVELS

The National Bureau of Standards has (among other things) the responsibility for maintaining *primary standards* that enable comparison to be made with the unvarying physical quantities. For instance, time and frequency are referenced to the cesium-beam standard (''atomic clock''), an oscillator regulated by the quantum mechanics of the cesium atom.

Government and other agencies at the next lower level have calibration centers that provide a link between the NBS primary standards and the *working standards* that calibrate test equipment used at workshop or flight-line level. These intermediate standards are called *secondary standards*. Periodically, they are sent to NBS for recalibration against primary standards. Less accurate than NBS standards, they are nevertheless extremely sensitive instruments that must be maintained in a special environment and handled with great care.

At shop level, the working standards that are used to perform periodical calibration of test equipment used in engineering research or product quality control go at regular intervals to be recalibrated against the secondary standards.

In this way, performance or other data measured during proof-testing of factory products, servicing of radio equipment, and so on, are traceable back to the basic physical quantities via a test equipment ''pyramid'' in which each level is more accurate than the one beneath it, as in Figure 1.2.

We shall go into this further in a later chapter. For the present, just note that traceability to a common yardstick results in standardization of all measurements, just as clocks are regulated to standard time. The accuracy of the measurement gets less as you descend from level to level, which means that the uncertainty of the measurement grows wider. Ideally, each level should be ten times more accurate than the one below

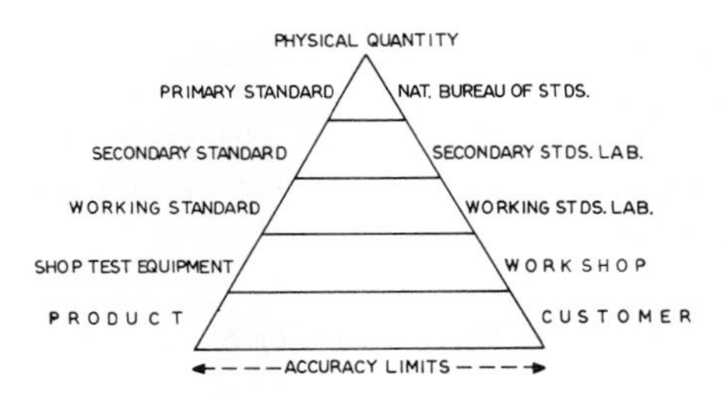

Figure 1.2 Test Equipment Pyramid

it. However, you should always buy equipment commensurate with the cost and purpose rather than get the most expensive on the market solely because it is the most accurate. A ratio of 4:1 or 5:1 is considered acceptable in modern practice.

If you have to use an instrument with an accuracy that is not even that much better than the tolerance allowed for the measurement, you must make allowance for it by "shading" the permissible limits. This means you must subtract the instrument's percentage accuracy from the tolerance allowed to obtain new (closer) limits.

To visualize what this means, look at Figure 1.3. Suppose you are measuring the value of a resistor. 0 is the nominal value marked on it. However, this value has a tolerance of ± 5 percent, so that any value not more than 5 percent above or below the nominal will be acceptable. OX represents these limits.

If your ohmmeter had no error at all, and it indicated that the resistor was at the permissible limit of 5 percent, you could accept that. But if your ohmmeter's accuracy is also 5 percent, than there is an uncertainty of ± 5 percent in the measured value, so that the real value might be as much as 5 percent higher or lower than the meter's indication. In other words, all you know about the real value is that it lies between − 10% and + 10% of the nominal value. This is represented by AB in Figure 1.3. You can see that there is a fifty-fifty chance that your resistor is outside the allowable limits.

A more accurate meter would improve this somewhat by reducing the area of uncertainty. One with an accuracy of ± 2 percent, for example, would shorten the spread to ± 7 percent, but you could still be outside the limits, as shown by CD in the figure. To get around this you must cut the resistor's allowed tolerance by subtracting the meter's accuracy from it:

Permissible tolerance of measurement	±5%
Accuracy of measuring instrument	±2%
Shaded tolerance of measurement	±3%

In Figure 1.3 the shaded tolerance is represented by X^1. EF is the meter accuracy. The area of uncertainty due to the meter is now all within the allowable limit OX.

TRANSDUCERS

Since electronic test equipment measures the behavior of free electrons, it also can measure any other quantity that can be converted into

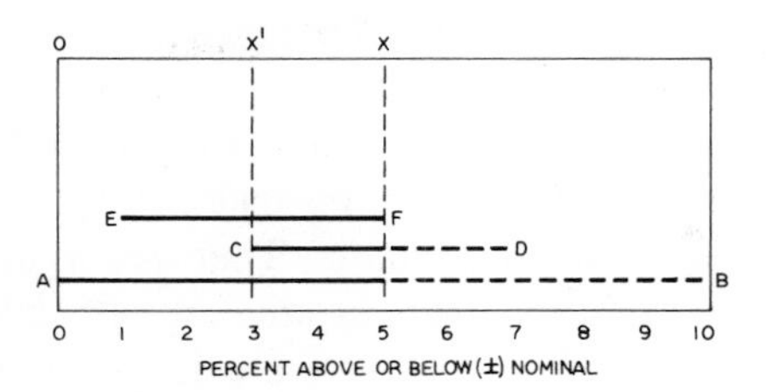

Figure 1.3 Shading Tolerance of Measurement

electricity. Any device that changes one kind of energy into another is a *transducer*. A photocell is a transducer that converts light into a current that can operate a meter. The strength of the current is proportional to the intensity of the light, and the deflection of the meter pointer is proportional to the strength of the current. The meter will not be calibrated to read the strength of the current, but the intensity of the illumination.

Table IV lists a number of transducers that enable electronic test equipment to measure other quantities by converting them into electrical energy. Since the amount of energy is usually very small, many of these instruments require an amplifier to obtain a useful indication. Such an amplifier is built in as part of the test equipment.

TABLE IV—TRANSDUCERS

Physical Quantity		*Transducer*
Heat (temperature)		thermocouple, thermistor
Light		photocell (photovoltaic, photoconductive)
Magnetism		Hall-effect probe
Mechanical:	acceleration	accelerometer
	angular displacement	"synchro" transmitter
	flow	flow sensor
	force	load cell
	level	level detector
	linear displacement	linear variable differential transmitter
	pressure	pressure transducer
	strain	strain gage transducer
	torque	torque transducer
Radio frequency		antenna
Sound:	audible	microphone
	ultrasonic	piezoelectric sensor

AMPLIFIERS

At the beginning of this chapter you saw that in many cases only a small sample of electron energy can be taken without affecting circuit performance or measurement accuracy. There may also be very little energy in the first place. Transducers produce very small signals in many cases. To get a useful reading on the indicating device (meter, digital readout, cathode-ray tube) an amplifier is necessary. The basic requirements of an amplifier, therefore, are:

(1) to boost weak signals;
(2) to avoid disturbing the signal source.

Amplifiers may be separate instruments used in conjunction with other test equipment, or they may be built in. They operate by introducing external power, which is controlled by the input signal in such a way that the output from the amplifier is an enlarged version of the input. No one amplifier is suitable for every application, of course, so there are many different types. These fall into two main groups:

(1) AC amplifiers;
(2) DC amplifiers.

AC Amplifiers

Ideally, an AC amplifier should have a flat frequency response and not distort the waveform of the signal. This is usually achieved by a large negative feedback, which also gives increased stability of measurement by compensating for changes in tube or transistor performance.

AC amplifiers may be wide-band, narrow-band or differential (see below).

DC Amplifiers

Ordinary DC amplifiers are not as stable as AC amplifiers; consequently, for more precision, many instruments convert a DC input to AC, and then amplify it with an AC amplifier. Conversion may be done by mechanical means with a chopper, or electronically with a transistor. A third method is to use photoconductors controlled by neon tubes in a relaxation oscillator circuit. In each case, the modulating circuit chops the DC into pulses with amplitude proportional to the DC input.

Differential Amplifiers

A differential amplifier is really a pair of identical amplifiers, each with its own input (see Figure 9.11). If you apply the same signal to each input, there will be no output because the two signals will cancel each other. However, if the two input signals are not identical the amplifier will amplify the *difference* between them.

Such amplifiers can be used to measure the difference between two signals. They can also be used to eliminate unwanted signals, such as hum or other interference picked up by both inputs. These unwanted signals are called *common-mode* signals, and the ability of a differential amplifier to reject them is called *common-mode rejection*.

A single-ended or unbalanced input may be applied to both halves of a differential amplifier by means of a phase-splitter, as in other push-pull amplifiers. In this case, the output represents the *sum* of the inputs, since they are of opposite phase.

ENVIRONMENTAL FACTORS AFFECTING ACCURACY

Temperature

Most test instruments can operate well outside the range of temperature people work in, but, under normal conditions, they will be used at around +20°C (68°F). However, laboratory test equipment, especially calibration standards, require an environment more closely controlled, because even quite small temperature variations affect the accuracy of measurements. Operation at higher-than-normal temperatures is not advisable for any test equipment.

Humidity

For general use a relative humidity not in excess of 90 or 95 percent is satisfactory. Above this level serious leakage will occur because of excessive dampness. Laboratory equipment has to be operated within narrower limits.

Barometric Pressure

Barometric pressure is seldom a problem at ground level, but may become so at altitude. Most test equipment will operate satisfactorily up to about 15,000 feet.

Vibration and Shock

Test equipment should be able to withstand normal handling and transportation, but should not be subjected to rough treatment.

Electromagnetic Interference (EMI)

Adequate shielding and grounding should be provided for all test equipment. Generally, the cabinet or dust cover of an instrument is metal, and will do a good job of excluding much unwanted interference. But this is true only if it is well grounded. The third pin of the power plug gives an electrical ground, necessary for safety reasons, but this is by no means perfect, since there are often considerable lengths of conduit and wire between the wall outlet and the real ground. Because of the presence of resistance along this path, especially in a dry climate, considerable unwanted voltage can be present on the chassis or case.

A real ground should be a copper rod or tube driven deeply into moist earth, or the equivalent, connected with thick copper cable by the shortest possible route to the chassis or cabinet. Domestic water pipes usually work well enough, if the run of pipe is not too long; but in a factory they often run for great distances overhead, and offer considerable resistance.

Much of this pickup will be in the form of hum, which is a low audio frequency having the same frequency as that of the power line or a harmonic thereof, introduced into the signal paths by induction, leakage or insufficient filtering.

Any electrical disturbance which causes undesirable responses in electronic equipment is called interference. This could be undesired signals, stray currents from electrical apparatus, or other causes such as "static" from atmospheric disturbances. A special case is noise, which is unwanted energy, usually of random character, present in any transmission channel or device and due to any cause. It may be due to the electrons themselves in the circuits under test, or connected to them, since the movement of each one is a tiny electronic current that may be amplified enormously in a powerful amplifier. The accumulation of millions of random electron movements creates the types of noise described in Chapter 7.

This unwanted interference competes with the signal you do want. The smaller the signal, the greater the problem. The ratio of the magnitude of the signal to that of the noise is the signal-to-noise ratio. This can be improved in many cases by a combination of interference reduction and avoidance of unnecessary signal attenuation, as discussed in the following paragraphs.

DECREASING INTERFERENCE

Decreasing the interference depends upon its source. There's not much you can do about noise originating in the test equipment itself, assuming it is working properly. But external interference can be greatly reduced by proper shielding and grounding of your test setup.

The manufacturer provides adequate shielding in his instrument, but you are responsible for the connecting leads and grounds. Good shielding in the instrument will be nullified by using unshielded or ungrounded equipment. An unshielded input picks up interference like an antenna and feeds it directly into the signal path at the point where it will receive maximum amplification. The same applies to a shielded cable, or the test instrument itself, if the shield, chassis or cabinet is not properly grounded.

This is illustrated in Figure 1.4. In (a) an unbalanced connection is used, with only one signal path. The return connection is via the shield.

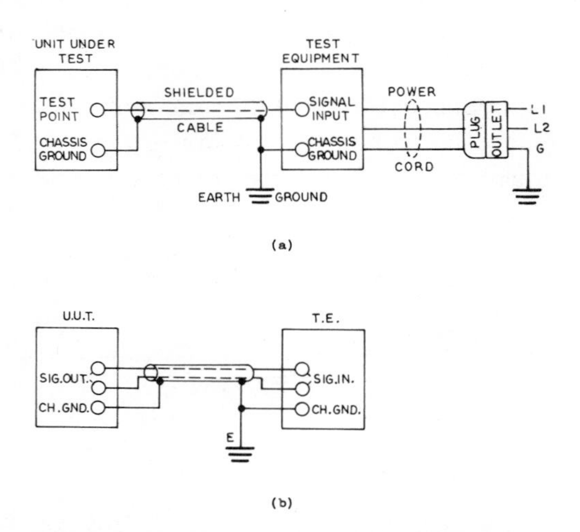

Figure 1.4 Proper Shielding and Grounding

The unit under test, the test equipment and the cable shield are all grounded. The expression "chassis ground" in this case means the low or zero-potential side of the circuit. The power cord of the test equipment (and of the unit under test, if applicable) is a three-wire cable terminated in a three-pin plug, plugged into a corresponding service outlet. As already explained, this third pin should be regarded more as a safety device than as a proper ground. In fact, it is sometimes desirable not to use it in very sensitive measuring equipment if it feeds noise in!

In recent years, many instruments that are battery-powered have appeared, combining greater portability with complete isolation from the power line. This is illustrated in Figure 1.4 (b), which also illustrates the connections for a balanced connection with two signal paths. In this case, both leads must be shielded, of course.

AVOIDANCE OF SIGNAL ATTENUATION

Apart from the question of loading the circuit, which we discussed at the beginning of this chapter, you have to ensure that connections between your test instrument, the unit under test and other test equipment do not attenuate or distort the signal (except in those cases where you actually want to do so, of course).

Loss of signal strength can be caused by impedance mismatches. This occurs when the output impedance of the circuit under test is not the same as the connecting-cable impedance or the input impedance of the test instrument. An impedance mismatch may also introduce distortion and phase shift. Of course, this does not apply to DC connections, and is very much worse at high frequencies than at low ones. Dissimilar impedances should be connected by using a *matching pad*.

For many measurements you should use a probe. A probe can avoid loading and mismatches. Some specimens are shown in Figure 1.5.

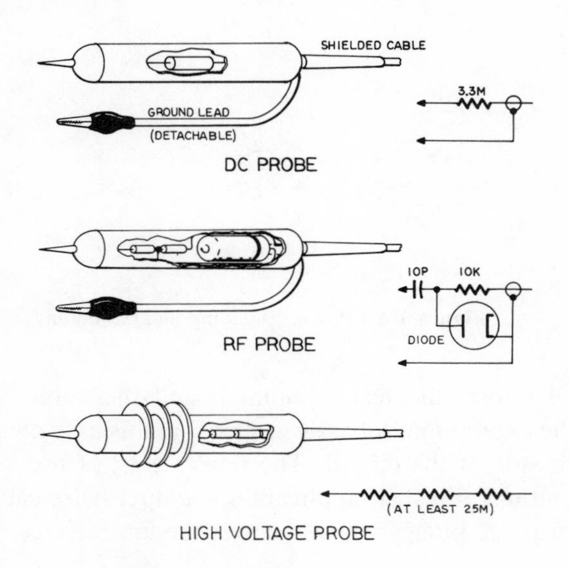

Figure 1.5 Probes

Probes frequently attenuate the signal by fixed amounts, such as X10 or X100, which means that the signal is decreased by these factors. Such reductions are necessary with signal voltages higher than what the test equipment can withstand, and also to increase the input impedance of some instruments to prevent loading. Many probes also contain adjustments whereby they can be matched exactly to the instrument with which they are used. You'll read more about this in Chapter 11 in connection with oscilloscope probes.

VECTORS

Anything that has the property of being measurable in dimensions, amounts, etc., is a *quantity*. Some quantities, such as mass, density and temperature, which are completely determined by their magnitude alone, are called *scalars*. Other quantities, such as velocity, acceleration and force, which have direction as well as magnitude, are termed *vectors*.

A vector quantity can be represented by an arrow pointing in the direction in which the force acts, and having a length proportionate to its magnitude. A simple example of this would be a wind velocity of 25 knots from the west. Using a scale of 10 knots to a centimetre, we could represent this with an arrow AB pointing to the east (the direction toward which the wind is blowing), and with a length of 2.5 centimetres.

WEST A ——————→ B EAST

2.5 cm

The wind is the movement of a mass of air, and anything floating in it, such as a cloud or a balloon, which is not attached to the ground and does not have a separate velocity of its own, will have the same velocity as the wind, and therefore the same vector represented by the arrow AB. However, if it should have a separate velocity of its own, such as an airplane would have, two forces will be acting on it: the wind and the thrust of the airplane's engine. These two forces combine to act together as if they were a single force, called the *resultant*. The simplest example of this is seen when the airplane is flying in the same direction as the wind—that is to say, it has a tail wind of 25 knots. The vector of the airplane's own velocity (100 knots to the east) is represented by the arrow BC, which is therefore drawn in an easterly direction with a length of 10.0 centimetres to represent 100 knots. If the air mass were stationary, the thrust of the engine would be the only force acting on the airplane, so it would have this vector alone. But as the air mass is moving at 25 knots in the same direction, its vector AB is also acting on the airplane, so that the plane's total velocity relative to the ground is given by AB + BC = AC, or 25 + 100 = 125 knots. AC is, therefore, the *resultant* of the components AB and BC. Finding their resultant is called the *composition of forces*.

WEST A——————→——————→ C EAST
2.5 cm B 10.0 cm

If the airplane turns around and flies in exactly the opposite direction its tail wind becomes a head wind. The arrow BC, representing the airplane's own separate velocity, is now drawn pointing toward the west. You can think of the arrow being the force of the propeller pulling the airplane at B toward C. But the force of the wind is also acting on the airplane at B as before, except that these two vectors are now acting in opposite directions, or AC = BC − AB = 100 − 25 = 75 knots.

WEST C ←——————●——————→ B EAST
7.5 cm A 2.5 cm

Summarizing this, then, we say that component vectors acting in the same direction are added to obtain their resultant; and if they act in opposite directions they are subtracted. But what happens if they are not acting in the same direction or its opposite?

Suppose, for instance, that the wind is blowing from the southwest at 25 knots when the airplane is headed east at 100 knots. As before, the airplane's vector is represented by the arrow BC, its length indicating the speed at which the propeller is pulling it eastward through the atmosphere. But we must also indicate the effect of the wind, which is represented by BA, a force acting to move the airplane at 25 knots toward the northeast. This is not exactly a tail wind, but obviously it will increase the airplane's groundspeed somewhat.

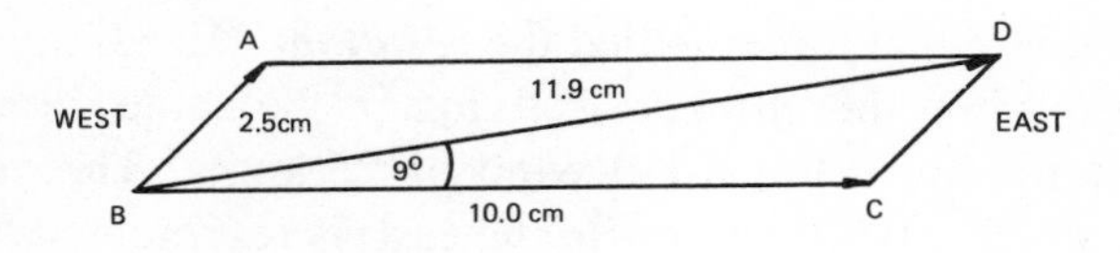

To find out exactly what happens we must determine the resultant of these components, BC and BA. To do this, we draw AD parallel to BC and CD parallel to AB. This gives us a parallelogram ABCD, known as the Parallelogram of Forces. The diagonal BD is the resultant.

This resultant is 11.9 centimetres long, which tells us that the airplane's groundspeed is 119 knots. Also, the direction of BD is not the same as that of BC. Because the wind, from the pilot's point of view, was blowing from the right, the airplane was drifting to the left. The angle of drift, DBC, is 9 degrees. The airplane will not reach its destination unless the pilot steers enough to the right to offset the effect of the wind.

The resultant BD can be *resolved* into two other component forces acting on the airplane: BX in the direction it is headed, BY at a right angle to it.

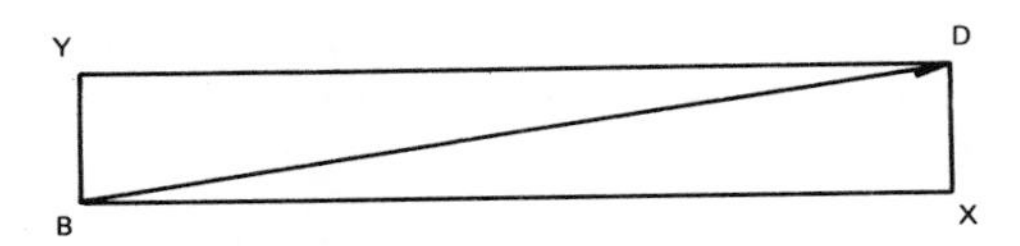

These are not the airplane and wind velocities, of course, but are two artificial vectors that can be plotted on a piece of squared paper. Although not of much interest to the air navigator, this graphical representation of the composition of two forces acting perpendicular to each other is used in physical science to analyze many functions. In electronics it is most commonly used to resolve an impedance into its components of resistance and reactance.

When an alternating current flows through a "non-inductive" resistor, the variations in amplitude of current and voltage remain in phase. But if the same current flows through a pure inductance, the current lags the voltage by a phase angle of 90 degrees. If an alternating current is applied across a pure capacitance, we get the opposite result, as the current leads the voltage by 90 degrees. In each of these statements we emphasized a "pure" resistance, inductance or capacitance, although such a thing does not exist in reality. A coil of wire has some resistance and capacitance as well as inductance; a capacitor has resistance and inductance in addition to capacitance. These may be shown geometrically in a rectangular coordinate system. For example, suppose you had a circuit that at some frequency had an inductive reactance of 500 ohms, a capacitive reactance of 50 ohms, and a DC resistance of 100 ohms. Through this circuit flows an alternating current of 1 ampere. In the diagram this current vector is assumed to lie in OX.

Taking the DC resistance first, we have 1 ampere flowing through 100 ohms. Since $E = IR$, the voltage across the resistance is 100 volts. We plot this along OX as OR because it is in phase with the current.

Each current peak produces a voltage of 500 volts across the inductive reactance, but as this leads the current by 90 degrees it is plotted on OY as OX_L. This is 90 degrees ahead of OX, assuming a counterclockwise rotation. Similarly, we plot a potential drop of 50 volts across the capacitive reactance as OXc along OY′, because it comes 90 degrees before the current peak.

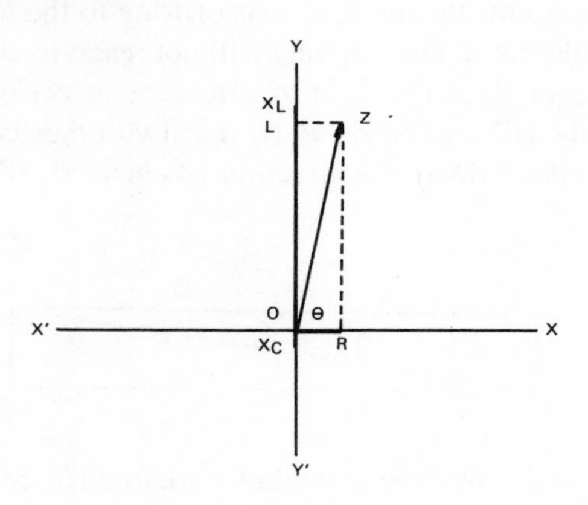

The vectors OXL and OXc are in the same straight line, but acting in opposite directions. OXL is positive and OXc is negative, therefore, the resultant of these two forces will be OXL—OXc, or OL. OL is, therefore, the total reactive force to be considered.

Now we draw a line, LZ, parallel to OR, and another, RZ, parallel to OL. This gives the parallelogram of forces from which we can obtain the resultant OZ of the vectors OL and OR. This resultant is the actual voltage peak that would be measured in this circuit, and is the voltage drop across the impedance, or AC resistance, of the circuit. Measuring OZ we find it is 461 ohms. We could also calculate this from

$$OZ = \sqrt{OR^2 + OL^2}$$

The angle θ is the actual phase angle of the circuit. It is 77 degrees; less than 90 degrees because of the resistive element. It can also be calculated from $\tan\theta = \frac{OL}{OR}$.

We shall have more to say about impedance in Chapter 4, and again in Chapter 11, where we also mention *complex numbers*. The use of complex numbers is just another way of stating the component vectors of a quantity, such as impedance, instead of drawing them. For example, we can say that a given impedance Z is equal to the square root of the sum of the squares of its component vectors, or:

$$Z = \sqrt{R^2 + X^2}$$

We can also write it: $Z = R + jX$ where R is the value of the resistive component and $+X$ that of the inductive (a capacitive reactance would be $-X$). The "operator" j means that the vector X is perpendicular to R, and therefore cannot be added or subtracted directly. The two values are consequently coordinates that define the position of Z on a graph, as we've just seen.

j is written i when used for non-electronic purposes, but in electronics i is already used to symbolize alternating current. Its use was introduced to solve the equation $x^2 + 1 = 0$, which becomes $x^2 = -1$ when we transpose $+ 1$ to the other side. To find the value of x we have to find the answer to $x = \sqrt{-1}$. The expression i (or j in electronics) is used instead of $\sqrt{-1}$, but don't try substituting it for j in these expressions as it will only cause confusion. Complex numbers are very much a part of the scientific world. There is no need, however, to discuss them further in this book.

DECIMAL NUMBER SYSTEM

Primitive man must have used his fingers to indicate small numbers, just as children do today; hence we use ten *digits* (fingers) in our decimal number system.

The decimal number system resembles your automobile odometer. Each position can be occupied by any one of the numbers 0 through 9 printed around the edge of the little wheels in Figure 1.6 (a). The value indicated depends on which position the number is in, so this number system is called *positional notation*.

In the decimal system there are ten numbers, so it has a *base* of ten, and each position is a power of ten, starting with 10^0 and going to the left. Values less than 1 go to the right of the decimal point as if they were a mirror reflection of those to the left.

BINARY NUMBER SYSTEM

There is nothing sacred about the decimal number system. Number systems with other bases are equally valid; it's just a matter of convenience. The binary number system has a base of two.

This works better for many purposes, since each position (power) has only *two* numbers, 0 and 1. If your odometer were on the binary system, it could have a light at each position instead of a number wheel, and the lighted bulbs would indicate ones, the unlit bulbs zeros. For instance, if the bulbs lit in the pattern shown in Figure 1.6 (b) the number would be 6.

This number system is therefore very useful in electronics, where

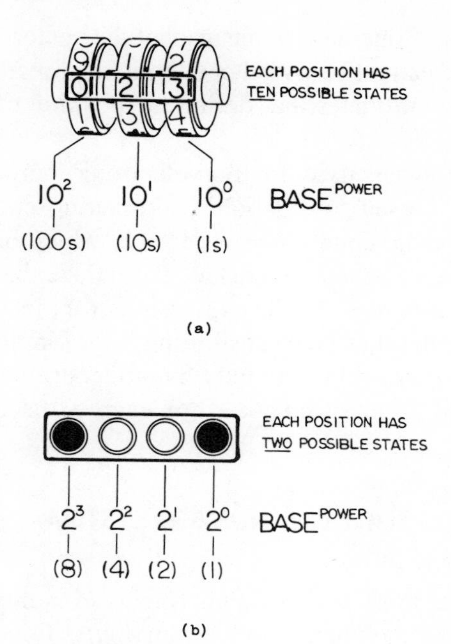

Figure 1.6 Number Systems

simple on-or-off states abound. Decimal numbers can be converted into binary, which as voltages can be manipulated and, if required, can then be reconverted to decimal. Computers operate on this principle, both in performing calculations and in issuing control instructions to other equipment. Table V gives the binary equivalents of the digits 0 through 9.

TABLE V—DECIMAL/BINARY CONVERSION

Decimal	*Binary*	*Decimal*	*Binary*
0	0 0 0 0	5	0 1 0 1
1	0 0 0 1	6	0 1 1 0
2	0 0 1 0	7	0 1 1 1
3	0 0 1 1	8	1 0 0 0
4	0 1 0 0	9	1 0 0 1

Figure 1.9 illustrates the conversion of decimal numbers to binary, and one use of binary: in punching a tape, where holes are punched for ones. In this way, the decimal number 269320 can be stored in binary form on the tape. Later, it can be recovered by running the tape through a tape reader, in which each hole causes an electrical pulse in a wire corresponding to its position, these pulses being used to operate a printer printing decimal figures.

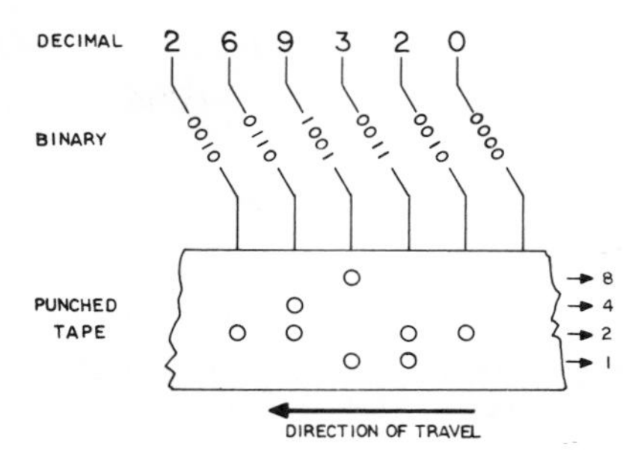

Figure 1.7 Decimal/Binary Conversion

SUMMARY

In this chapter you've seen what electronic test equipment is and does. You have been introduced to many of the factors affecting its accuracy, and been shown how any measurement can be traced back to its prototype physical standard. You have also reviewed vectors and binary numbers, which will be of help to you in understanding some of the test equipment described later on.

2

UNDERSTANDING DIRECT READING DEVICES

Electronic test equipment can be divided into *electronic instruments* that won't work without external power and *direct-reading devices* that draw such power as they require from the circuit being tested. In this chapter, we are going to discuss the second group, which embraces both simple indicators and a wide variety of meters. Some of the meter movements covered here are also built into more complicated electronic instruments, as you'll see in later chapters.

SIMPLE INDICATORS

Neon Tester (Figure 2.1)

At the beginning of Chapter 1, the neon tester was cited as an example of a simple indicator that converts invisible electric current into visible light. This is an inexpensive device used by electricians to find out if a circuit is ''live.'' It consists of a neon bulb in an insulated holder with two leads terminated by phone tips. A current-limiting resistor (100 kΩ) is installed in series with the bulb to prolong its life.

Figure 2.1 Neon Tester

To use the neon tester, one of the two leads is connected to the ground or low side of the circuit, or just held in the hand. The other lead is touched to the ''hot'' terminal or conductor. If voltage is present, the gas in the bulb ionizes and glows.

The tester works on either AC or DC. It indicates AC by illuminating both elements, DC by illuminating only one.

Pilot Lights

A pilot light can be any small lamp used as an indicator. Most pilot lights show whether something is ''on'' or ''off,'' usually the main supply voltage to a piece of equipment, in which case the lamp is placed close to the power switch. Other uses are to warn or inform the operator of something that he should be aware of. A familiar example is the ''idiot light'' on an automobile dashboard that lights when the ignition is switched on, but goes out again when the engine starts. If it comes on when the engine is running, it means that the oil pressure is too low. Another example is the brake light that reminds you that the hand brake is on.

Pilot lights use either neon or incandescent bulbs. The neon has the advantages of long life and negligible current consumption, but the incandescent bulb is brighter. The most popular colors for the lenses of these lamps are red, green, blue, amber or yellow, white or clear. (Green, blue and white do not work well with neon bulbs, however.) Some lights are constructed with a *legend plate,* which is an oblong piece of translucent plastic bearing a word or message that appears when the lamp lights (''placard lights'').

In some applications, a push-button switch is combined with the lamp so that the switch button lights when the switch is on. In another version, a built-in switch allows you to test the lamp's own bulb by pressing the lens (''press-to-test'').

Incandescent bulbs are available for voltages from 1.3V to 250V. Neon bulbs are manufactured for 55V, 105-125V and 220-250V, and are frequently mounted in holders with built-in current-limiting resistors. Both types of bulbs may have bayonet or screw-type bases, or plain leads for permanent installation.

Audible Devices

Very little use is made of the sense of hearing in electronic testing, because although the ears are very sensitive, they cannot read numerical indications as the eyes can. However, they can hear sounds from all directions at once, whereas the eyes can see only what they are looking at, so an audible warning device is often attached to test equipment to draw attention to conditions which then can be evaluated visually. This is most likely to be done in large complexes of test equipment, where important details (such as temperature) might go unnoticed for some time because the operator is attending to something else.

This type of audible warning device usually takes the form of a relay-operated horn or buzzer. A sensor of some kind detects the critical condition and originates a signal that, after amplification, actuates the warning device.

Although your ears cannot tell you everything your eyes can, they can distinguish very low levels of sound. This makes it possible to use a pair of headphones instead of a galvanometer when ''nulling'' a bridge (see Chapter 4), provided the AC energizing the bridge is within the audio range. As you approach the null, the sound gets fainter and then dies out. With further adjustment, the sound begins to come back. The soundless spot between where the sound died out and where it came back is narrow enough to give a sharp null.

METERS

Meters are electromechanical devices that transform invisible electrical energy into a visible mechanical indication. This indication usually takes the form of the movement along a dial scale calibrated according to the quantity being measured, the travel of the pointer being proportionate to the electrical energy. Figure 2.2 illustrates typical modern panel meters.

Figure 2.2 Panel Meters *(Courtesy Triplett Corporation)*

The link between the electrical phenomenon and the mechanical response is the *meter movement*. Meter movements are either magnetic or non-magnetic.

PERMANENT-MAGNET MOVING-COIL METER MOVEMENT

The permanent-magnet moving-coil (PMMC) meter movement, or d'Arsonval meter movement, is the most commonly-used direct-current operated movement. As you can see in Figure 2.3, it consists of a small coil of wire wound on a light aluminum frame, supported on jeweled bearings between the poles of a permanent magnet. Spiral hairsprings hold the coil so that its attached pointer is at zero on the scale. These hairsprings are connected to opposite ends of the coil and conduct current to it.

When a current flows through the coil its magnetic field interacts with that of the permanent magnet, so that a turning force is applied to the coil. The coil rotates until the opposition of the hairsprings just balances the magnetic torque. The pointer now indicates the corresponding value on the dial scale. When the current ceases to flow in the coil the hairsprings return the coil and pointer to the zero position.

The iron core within the coil is stationary. Its function is to maintain a uniform field in the space through which the coil rotates so that its motion will be linear.

Occasionally, it is necessary to reposition the pointer over the zero mark. This is done by an adjustment on the front of the meter just below the dial that resembles a screwhead, as you can see in Figure 2.2. Turning this adjustment with a small screwdriver rotates a peg that is located between the Y-shaped projections mounted on the support bracket shown in Figure 2.3. This peg is offset from the axis of the screw, so that it turns

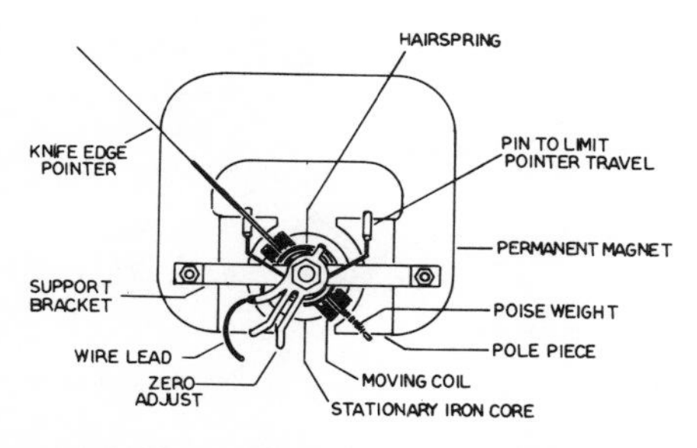

Figure 2.3 PMMC Meter Movement

in a circle, moving the Y-shaped projections first one way, then the other. One end of the front hairspring is attached to the opposite end of this adjustment, so that turning it changes the tension of the spring slightly, causing the pointer to shift its resting position.

This type of PM meter movement is comparatively inexpensive, rugged and reliable, and is used more than any other type. However, refinements have been introduced to overcome certain disadvantages, such as *pivot roll* and *pivot friction*. (The combination of these is called Friction Error.) Pivot roll is the condition where the movement is operated with the pivots in a horizontal position. Play in the bearings will allow one pivot to ride up in its jewel while the other slides down. This results in a tilt which introduces inaccuracy in the reading, and which will be different with each reading ("random error"). However, if the bearings are tightened to eliminate pivot roll you increase pivot friction.

One solution to this problem is to operate the meter with the pivots in a vertical position only, so that the lower one remains centered in its jewel. However, this is impractical in many cases. Two other improvements have therefore been developed, and will be found in the better models.

One of these consists in providing spring-backed jewels, so a slight inward pressure is applied to keep the tips of the pivots centered regardless of attitude. The tendency to increased pivot friction is countered by using highly-polished jewels.

The second method is to do away with jewels, pivots and hairsprings altogether, and replace them with thin metal strips held taut by tension springs. The movement is not pivoted but suspended by the metal strips, which twist as the meter turns, thereby applying the restoring torque. This is called *taut-band suspension*.

Most meters are constructed to provide for rotation over a 90° scale. However, long-scale instruments are also available with a scale length of 250°. This requires a different internal design of the magnet and core, as shown in Figure 2.4.

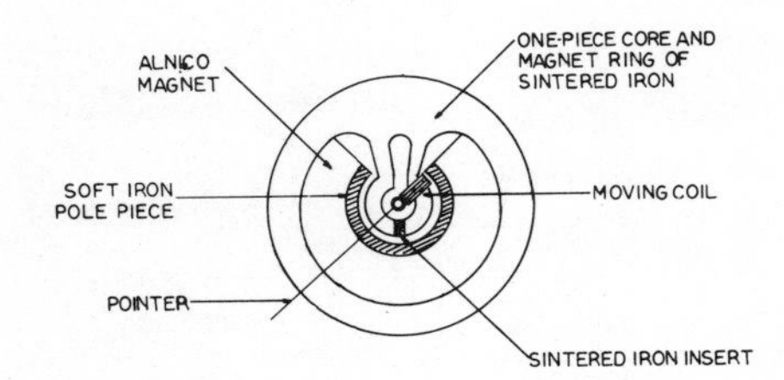

Figure 2.4 Long-Scale Meter Movement

Some other variations of the PMMC meter are described further on under meter applications.

IRON-VANE MOVEMENT

The iron-vane movement is the AC counterpart of the PMMC movement. It consists of two thin, soft iron plates ("vanes"), one fixed, the other pivoted, positioned within a stationary coil, as in Figure 2.5. When alternating current flows in the coil, the two vanes become magnetized with the same polarity, and repel each other. Springs, pivots and bearings are similar to those in the PMMC movement.

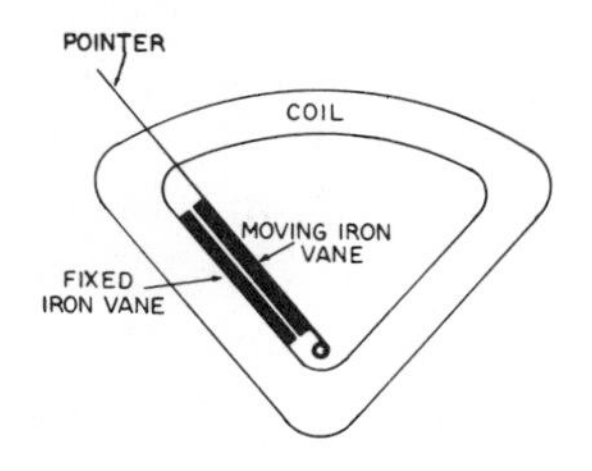

Figure 2.5 Iron-Vane Movement

The iron-vane movement is rugged and reliable, like the PMMC movement, and also inexpensive. However, the version shown in Figure 2.5 has a logarithmic deflection in which the lower end of the scale is compressed. To obtain a linear deflection the two vanes are made curved and mounted concentrically, the moving one in the form of a half-cylinder (180°), the fixed one a three-quarter cylinder (270°).

This type of movement is restricted to the frequency for which it is designed, which is necessarily a low one because of the iron.

ELECTRODYNAMOMETER MOVEMENT

The electrodynamometer movement consists of a moving coil mechanism like that of the PMMC meter, but instead of a permanent magnet, two or more fixed coils are used, as in Figure 2.6. This movement, therefore, resembles a cross between the PMMC meter and the iron-vane meter, and in fact can be used for both AC and DC.

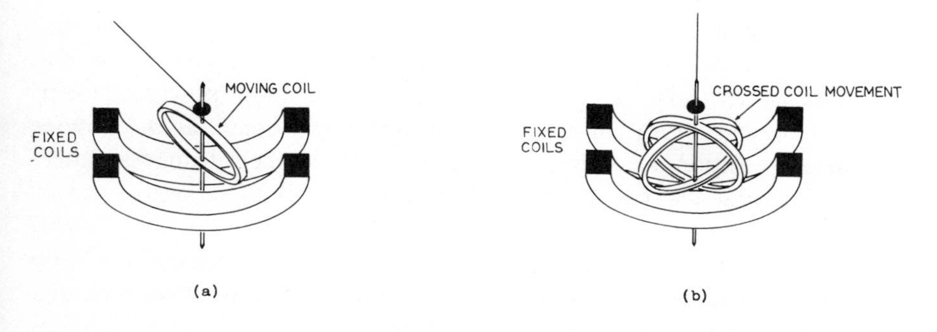

Figure 2.6 Electrodynamometer Movements

The same current flows in the fixed coils as flows in the moving coil; consequently, the magnetic field from the fixed coils varies also, and is weaker than that of the permanent magnet in the PMMC movement. This means that this movement is less sensitive than the PMMC movement.

Various arrangements of the coils adapt the electrodynamometer movement for different purposes. For current and voltage measurements, the fixed and moving coils are in series as shown in Figure 2.7 (a). The arrangement in (b) is for a wattmeter, in which the magnetic field from the fixed coils is proportionate to the current, and that from the moving coil is proportionate to the voltage across the load. (A resistor in series with the moving coil is used to change a current meter to a voltmeter, as explained further on under applications.)

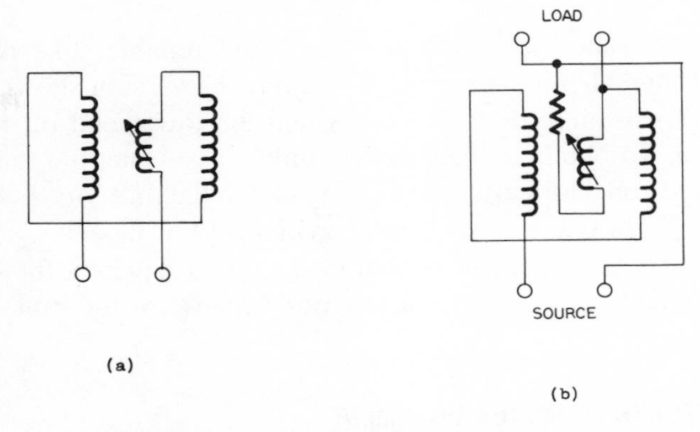

Figure 2.7 Electrodynamometer Circuits

The arrangement in Figure 2.6 (a) can be doubled by having two similar movements operating one pointer on a common pivot. This adapts the wattmeter for use in polyphase systems (three- and four-wire circuits).

Another arrangement has two moving coils mounted at 90° to each other within the two field coils, as in (b). This is called a crossed-coil movement, and it is used for measuring power factor. One coil is connected to a phase-shifting network so that it measures reactive power, the other measures resistive power (you'll read more about power factor in Chapter 4). No return springs are used in this movement as the pointer assumes a position according to the ratio the fields bear to one another. This movement also can be doubled to measure polyphase system power factors.

Another use for the crossed-coil movement is to measure the ratio between a standard capacitor and another capacitor to determine the capacitance of the latter.

Yet another version of the electrodynamometer movement has two pairs of crossed *field* coils. Instead of a moving coil, this meter has an iron vane. The fixed coils are connected to the source through inductive and capacitive elements. The position assumed by the iron vane, which is free to rotate in any direction, is according to the frequency of the source.

HOT-WIRE AMMETER MOVEMENT

The meter movements we have just described all depend upon magnetic fields for the conversion of electrical energy to mechanical. One disadvantage of this is that they can operate only at fairly low frequencies. In the hot-wire ammeter, we meet a movement in which electrical energy is converted to thermal energy in order to obtain the mechanical movement of a pointer.

In Figure 2.8, a length of resistance wire is stretched between A and B. When a current flows through this wire, it heats and expands lengthwise. A second wire attached to the first is kept taut by a spring anchored at C. When this wire moves, it causes the pointer to move. The expansion and contraction of the wire AB results in corresponding readings upon the dial scale. This scale is logarithmic, compressed at the lower end.

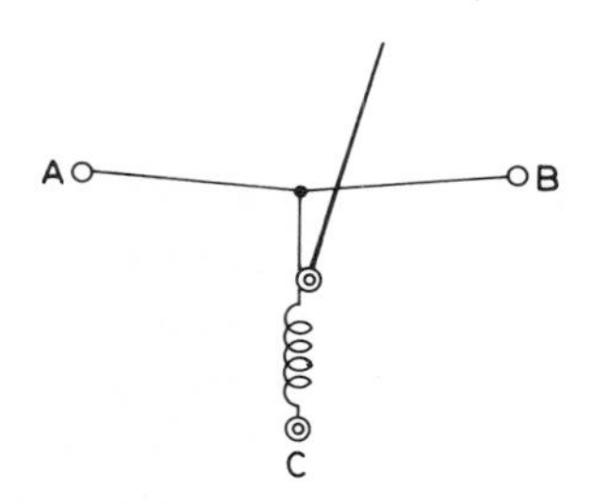

Figure 2.8 Hot-Wire Ammeter Movement

Hot-wire ammeters are not frequency-sensitive, and as they respond to the heat-producing capability of the current they don't mind whether it's DC or AC, or what type of waveform it has. However, they are rather sluggish in action because of the time it takes the wire to heat.

Electrostatic Voltmeter

The electrostatic voltmeter uses the electrical repulsion between two similarly charged metal plates to provide for the mechanical deflection of its pointer. As you can see in Figure 2.9, the movement resembles a variable capacitor. The moving plates are mounted on pivots working in jeweled bearings, with hairsprings, as in a moving-coil movement. Since both fixed and moving plates are connected to the voltage to be measured, they are charged with the same polarity. They repel each other, causing the moving plates to rotate until their electrostatic torque is balanced by the restoring torque of the hairsprings. The attached pointer then indicates the corresponding voltage.

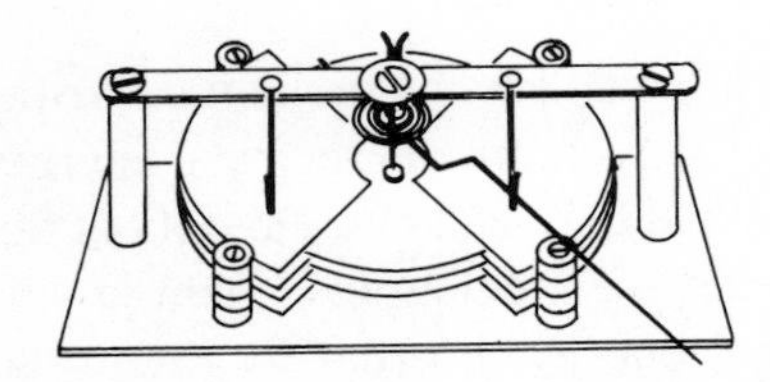

Figure 2.9 Electrostatic Voltmeter Movement

It takes considerable potential to create an electrostatic field strong enough to deflect the moving plates, consequently, this movement is used only in kilovoltmeters. The only current drawn is by leakage, but as the resistance to it is in the terohm range the loading effect of this instrument is negligible.

Gold-Leaf Electroscope

The gold-leaf electroscope works on the same principle as the electrostatic voltmeter. As you can see from Figure 2.10, it is a very simple device. When a charged object is placed in contact with the disk on top of the brass rod, the gold leaves at the other end, inside the sealed glass jar, separate because they receive the same charge, and therefore repel each other. The distance of their movement can be measured by a microscope with a calibrated scale which gives the magnitude of the potential. This instrument is more sensitive than the electrostatic voltmeter because of the extreme lightness of the gold leaves.

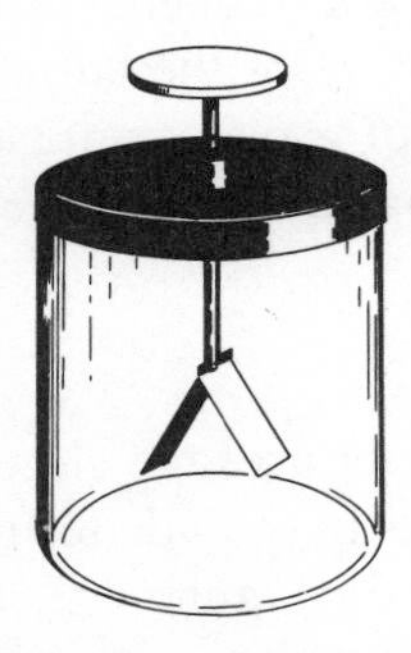

Figure 2.10 Gold-Leaf Electroscope

There is also a certain type of vacuum-tube voltmeter called an *electrometer* which is used for very small potentials, but as this is not a direct-reading instrument we shall defer discussing it until a later chapter.

METER CHARACTERISTICS

When you go to buy a meter, the manufacturer may provide you with specifications about the movement that look something like this:

ACCURACY	±1.0%
REPEATABILITY	0.5%
FRICTION	0.2%
DAMPING FACTOR	2
RESPONSE TIME	2 SEC MAX
OVERSHOOT	25%
TRACKING	±0.5%
TEMPERATURE INFLUENCE	±1%/10° FROM 25°C
SYMMETERY ERROR	±0.5%
FREQUENCY INFLUENCE	FLAT FROM 15 TO 10,000 HZ
POSITION INFLUENCE	1.0%

These are the principal characteristics of the meter. They are not always the same for every meter, of course, but vary somewhat depending upon the type. You ought to have some idea of what they mean, because they can tell you how suitable the instrument is for your purpose.

Accuracy is the measure of how close the meter reading is to the true value of the applied energy. It is usually given as a percentage of full scale. For example, if a meter with a scale reading from 0-100 milliamperes always reads between 99 and 101 mA when a 100-mA current is applied to it, its accuracy is ±1%. This meter will read within 1 mA of the true value at any point on the scale.

If the scale does not have zero at one end, the full-scale value will be the total value from one end of the scale to the other, regardless of sign.

Repeatability is the measure of the meter's ability to provide the same reading every time for the same input. The figure is a percentage of full scale, so on a 0-100V scale, for example, 0.5% would mean that the pointer would come to rest within half a volt of the same scale reading each time for the same input.

Friction means the maximum percentage of the full-scale value that the pointer may move if you tap it after it comes to rest. In this case, it may move not more than 0.2% of the full-scale value.

Damping Factor is a ratio obtained by a sudden application of current to a meter so that it deflects the meter to full scale before dropping back to a steady deflection. If the full-scale value is 100V, and the pointer drops back to 80V after the initial deflection, the damping factor is DF $= \frac{80}{100-80} = 4$. The higher the figure, the more heavily damped the movement is.

Damping is a means of controlling the swinging back and forth of the movement when a current or voltage is first applied. One way of doing this is to attach a light vane to the pivot. This vane is housed in a closed chamber so that the air in it acts as a brake, consequently, this method is called *air damping*. *Magnetic damping* is similar, except that the "brake" is a metal vane turning in a magnetic field. *Periodic damping* is where the movement oscillates about the reading before coming to rest; *aperiodic damping* is where it comes to rest without overshooting, sometimes called *overdamping*. If the pointer overshoots by less than half the rated accuracy of the instrument, it is *critically damped*.

Response Time is the time taken by the pointer to settle down after an abrupt change in the applied energy.

Overshoot is the amount by which the pointer overshoots when the applied energy is changed, and is expressed as a percentage of the deflection after the movement has come to rest.

Tracking means the ability of a meter to indicate accurately at each scale division. It is calculated by applying energy to give an exact full-scale deflection, after which the applied energy is reduced until the pointer is exactly over the next lower selected scale division, and so on down the scale. The actual values of applied energy for each deflection are noted. At any selected point on the scale, the actual applied energy will vary slightly from the theoretical value. For example, at the halfway point, the theoretical value of the applied energy should be half that required to give a full-scale deflection. The difference between this and the actual value, expressed as a percentage, gives the tracking error at that point. The manufacturer's figure will be the maximum error for any point checked.

Temperature Influence means the variation in meter readings due to temperature alone. In this case, it is expressed as within one percent for every ten degrees from 25° Celsius.

Symmetry applies only to *off-set zero* meters. These are meters in which the zero is at some place other than the end of the scale. Symmetry is the measure of the meter to give readings with equal accuracy on either side of the zero mark for equal inputs with opposite polarity.

Frequency Influence is a change in the reading due solely to a change in frequency of the applied energy. The expression "flat from 15 to 10,000 Hz" means that frequency influence is zero between these frequencies. If the change is due to a change in the waveform, it is called *waveform influence*. These two factors apply only to AC meters.

Position Error means the amount of change, if any, when the meter is operated in different positions—in the vertical instead of the horizontal, for instance.

METER APPLICATIONS

So far we have covered the construction and characteristics of meter movements. Now we can go on to the adaptation of these movements to various uses, and to the actual handling of meters.

Current Meters

A simple PMMC movement by itself is restricted to measuring DC currents from zero up to the maximum it can handle. This is not very large because the fine wire used in the moving coil cannot withstand more than a few milliamperes—in some cases only a few microamperes! In order to measure larger currents the excess must be bypassed around the movement by providing a *shunt resistor*, as shown in Figure 2.11.

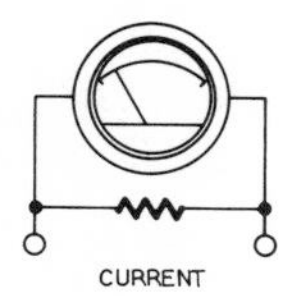

Figure 2.11 Current Meter with Shunt Resistor

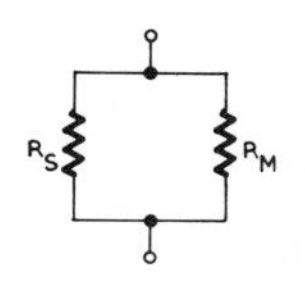

Figure 2.12 How a Meter Shunt Works

The value of this resistor depends upon the meter's *sensitivity* (the current that causes full-scale deflection), *resistance*, and current *range* to be measured. If you have a 500-microampere movement with a resistance of 1000 ohms, and you want to use it as a DC milliampere with a range of 0-1000 milliamperes, you must choose a shunt that divides the full-scale current of 1000 milliamperes so that only 500 microamperes go through the meter and the rest through the shunt.

This is shown in Figure 2.12, where RM is the meter resistance and Rs the shunt resistor. When 500 microamperes flow through the meter (RM), the potential drop across it is .5V ($E = 1 \times 10^3 \times 500 \times 10^{-6} = .5V$). Therefore Rs must be of such a value that the same potential applied to its results in a current of $1000 - .5 = 999.5$ milliamperes. Its value will be:

$$Rs = \frac{.5}{.9995} = .50025\Omega \text{ (approx.)}$$

DC Voltmeters

You can use the same movement as a voltmeter. If you want it to have a range of 0-100V you need a resistor in series with the meter, as in Figure 2.13, with a value such that the current produced by applying 100V across both will be 500 microamperes through the meter. The total resistance is:

$$R_T = \frac{100}{500 \times 10^{-6}} = 200k\Omega$$

The series resistor required will thus have a value 200 − 1 = 199kΩ. As this meter and shunt have a total resistance of 200kΩ at 100V, it has a loading effect of 2000 *ohms per volt*, a characteristic that may be important in some circuits, as you saw in Chapter 1.

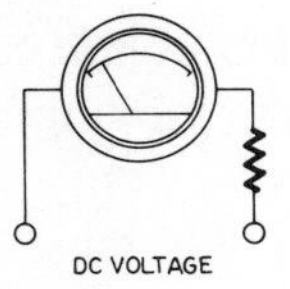

Figure 2.13 DC Voltmeter

AC Voltmeters

PMMC movements operate only on direct current. However, they are rugged and inexpensive, and are, therefore, used for AC more often than iron-vane and electrodynamometer movements. To convert a PMMC meter for AC you add a rectifier in series with it, as shown in Figure 2.14, or use it in a bridge circuit as in Figure 2.15.

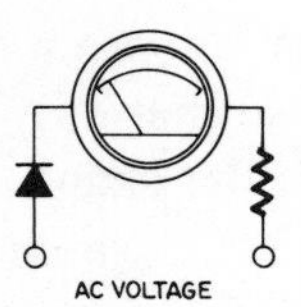

Figure 2.14 AC Voltmeter with Single Rectifier

In Figure 2.14, where the movement is used as an AC voltmeter, the rectifier permits only every half-cycle to pass. Deflection of the movement is always proportional to the *average value* of the rectified wave. However, you're more usually interested in the *RMS value*. The average value of a sine wave is 0.636 of the peak value, the RMS 0.707. By making the divisions of the scale worth 1.11 times the average value they will give the RMS value. But this will be true only for sine waves. On other waveforms, the pointer will also point to 1.11 times the average value, but this will *not* be the RMS value.

Full-wave rectifiers, as in Figure 2.15, develop twice the torque because both halves of the sine wave are used.

Figure 2.15 AC Voltmeter with Bridge Rectifier Circuit

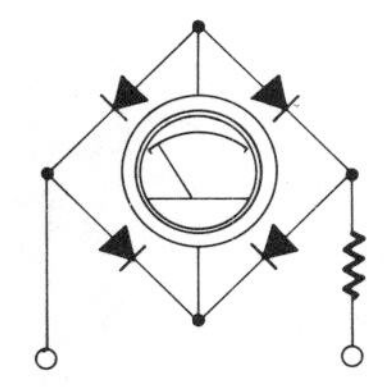

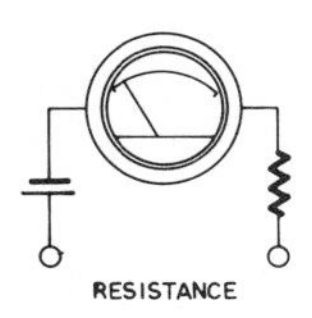

Figure 2.16 Ohmmeter

Ohmmeters

The PMMC movement is also used to measure resistance, as shown in Figure 2.16. The resistance to be measured is connected across the terminals and becomes part of the circuit with the meter, series resistance and battery. You will read more about ohmmeters in the next chapter.

Galvanometers

One of the factors limiting accuracy is *resolution*. This simply means the number of divisions on the scale. The larger the scale, the more divisions you can have, but then you need a bigger pointer, which in turn is limited by the torque of the movement. One way of getting a large pointer without increasing weight is to use a light beam. A ray of light from a lamp is reflected by a small mirror attached to the movement, to fall on the scale. The scale in this case is translucent, with the movement behind it. It lights up where the beam strikes it, the illuminated portion moving along the scale according to the applied energy. This type of movement is often used for *galvanometers* and similar meters requiring both high sensitivity and high resolution. The use of galvanometers with bridges is discussed in the next two chapters.

Thermocouples

Another way in which the PMMC movement can be used for AC is by use of a thermocouple. A typical thermocouple circuit is shown in Figure 2.17. The thermocouple element is mounted inside a glass envelope, which is evacuated (except where the current is in excess of one ampere, where the heavier heater wire does not need the protection). The current to be measured flows through the heater, a fine wire of high-resistance material. When the maximum current for which it is rated is flowing through it, the temperature of the heater wire will be about 300°C. The thermal junction, composed of two dissimilar metals, is imbedded in a glass or ceramic bead on the heater wire, but not in electrical contact with it. When the temperature of the junction reaches 300°C, the voltage output into an open circuit is about 10 or 12 millivolts.

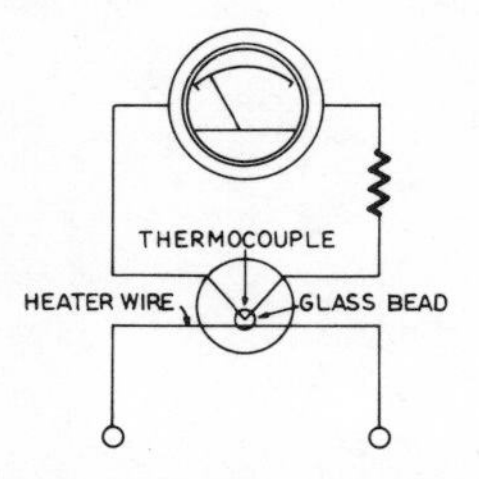

Figure 2.17 Thermocouple Meter

You must always use great care with thermocouple instruments. The temperature of the heater wire is proportionate to the square of the current. This means that if you double the heater current its temperature goes up four times. If it was already at the rated value, its temperature would shoot up from 300°C to 1200°C! Since the wire is very fine it can reach burn-out temperature while the meter pointer is still moving up the dial; after which the pointer drops back without ever reaching full scale, to the great indignation of the technician, who says: "I can't have burnt it out—it didn't even go all the way up!"

Apart from this disadvantage, however, thermocouple meters have a wide frequency range, true RMS response and great sensitivity.

Segmental Meters

Segmental meters or "expanded-scale" meters (Figure 2.18) are used primarily for monitoring AC line voltages. They are basically PMMC movements adapted for AC, but with an additional circuit. This circuit (a non-linear bridge) does not allow current to flow through the moving coil until the voltage reaches 100VAC. Between this voltage and 130VAC the meter reads normally. The accuracy is usually given as a percentage of center-scale value (115VAC) with a 60-hertz sine wave.

Meter Relays

Meter relays (Figure 2.19) are often used to maintain furnaces or environmental chambers between preset temperatures limits, and are usually PMMC movements with scales calibrated in degrees Fahrenheit or

Celsius. Two adjustable pointers are set to the upper and lower limits. When the temperature falls to the lower set point, the meter activates a relay that turns on the heater, turning it off again when the temperature reaches the upper set point; vice-versa for refrigerative units, of course.

The meter is actually responding to the current from a thermocouple. Switching may be done by having the pointer attached to the movement contact the set-point indicators. This causes a small current to flow, and triggers the relay circuit, which is usually solid-state. Alternatively, sensing is accomplished internally by an infinite-life lamp and photoconductors, the light from the lamp being interrupted by a vane attached to the movement. This type of relay is sometimes known as an optical meter relay. There are also others that operate inductively or magnetically.

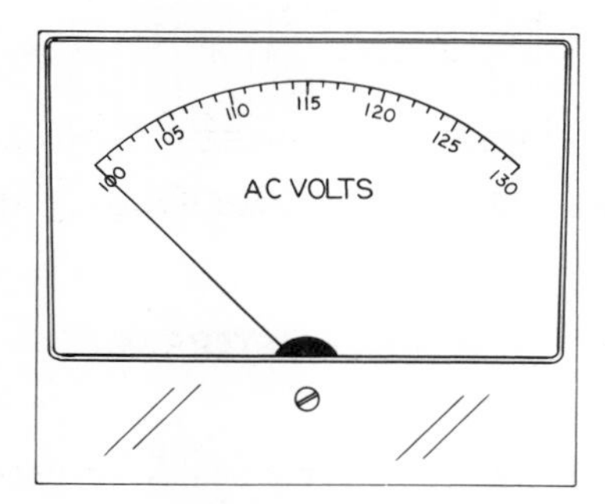

Figure 2.18 Segmental Meter

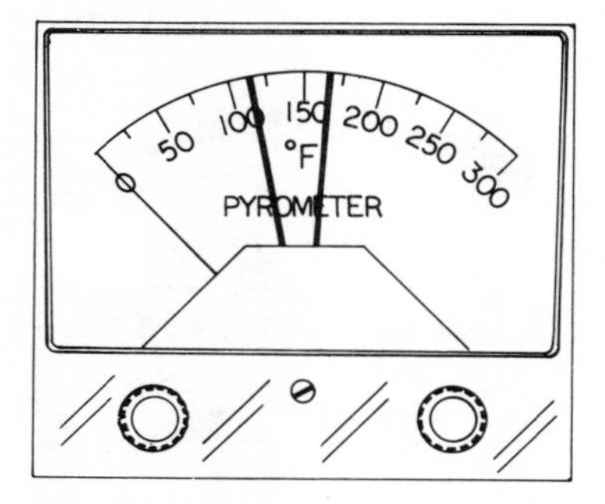

Figure 2.19 Meter Relay

Frequency Meters

Vibrating-reed frequency meters (Figure 2.20) are used to monitor AC-line frequency. This is most often 60 hertz, but many processes also require 400 hertz, so meters are available for both.

In this type of meter, five or more reeds are installed in a row. Each will vibrate at a certain frequency. In a 60-hertz instrument, there might be five reeds responding to 58, 59, 60, 61 and 62 hertz. In the one in Figure 2.20, there are nine, giving twice the resolution. Current flowing in a coil excites the reed tuned to its frequency, so that it vibrates. To the eye it seems to expand vertically. The frequency is, therefore, indicated by the reed that is "bigger" than the rest.

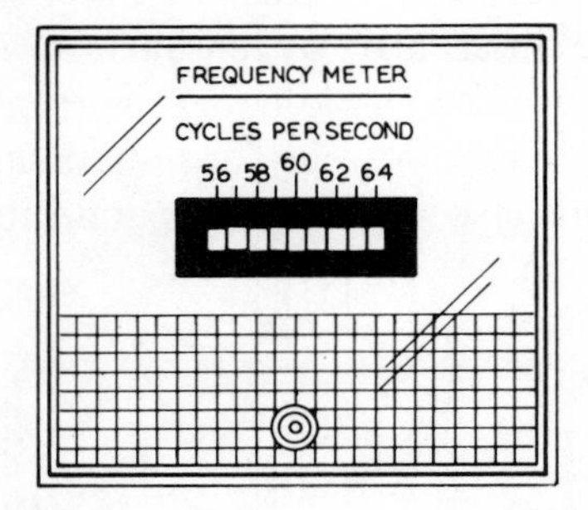

Figure 2.20 Vibrating-Reed Frequency Meter

USING METERS

Direct-reading meters are either portable or permanently wired into circuits and mounted on panels. Most portables are *multimeters,* in which one movement performs the functions of DC and AC voltmeters, ohmmeter and milliammeter according to the setting of the range switch. As

Figure 2.21 Portable Voltmeter *(Courtesy Weston Instruments, Inc.)*

multimeters are covered in Chapter 6, and resistance measurements discussed in Chapter 3, we shall here consider only the measurement of current and voltage using individual portable instruments, such as the one shown in Figure 2.21, although the principles are the same for any direct-reading meter.

Current Measurement

Portable meters for measuring current are ammeters if they are for currents of one ampere and up, or milliammeters for those below one ampere. They may have several binding posts, one of which is the common terminal; the others for different ranges. Alternatively, different ranges are selected by a range switch.

If you are measuring AC, make sure no DC can pass through the AC meter movement. If DC is present, connect a large capacitor, with a high-enough voltage rating, in series with the meter.

You should know approximately the value of the current you are going to measure, so that you can select a meter with the appropriate range. If you have no idea, a preliminary check with a multimeter is advisable, starting with the highest range and switching downwards until you reach the right one.

Having selected your meter, place it on the bench in its correct operating position. For many this is horizontal. Then check if the pointer is resting directly over the zero mark. If not, reset it, using the adjusting screw or knob on the front of the instrument. It is preferable to turn this so that the pointer is first shifted upscale, and then brought downscale to the zero mark. Tap the glass *lightly* while doing this to eliminate friction error.

These meters usually have a *mirror scale,* in which you can see the pointer's reflected image. Always keep the pointer in line with its image when taking a reading, to avoid *parallax error,* which is what happens when your eye is not directly over the pointer. Modern meters have unbreakable windows that have been specially treated against electrostatic effects. (When you wipe an untreated window clean, the resultant static charge attracts the pointer, giving you a false reading. This can be overcome by using a cloth moistened with water in which a little detergent has been dissolved.)

To measure current, you have to open the circuit at the point where the measurement is to be made, and connect the meter so that the circuit current flows through it. *Turn off all power before doing this.* If possible the connection should be made at the point with the lowest potential. Also, connect a short-circuiting switch across the meter terminals so that the current will bypass the meter except when you are taking a measurement.

Then turn the power on again, and allow sufficient warm-up time for the current to stabilize. Check the meter range before opening the switch, to avoid pegging the pointer. If in doubt, start with a higher range. Take the reading on a range that puts the pointer on the upper part of the scale, because this provides for greater accuracy.

The principal divisions on the scale are numbered with two or three sets of numbers corresponding to the ranges. Be sure you have the right one! Count the unnumbered increments carefully from the next lower-numbered division, and if the pointer stops in the space between the marks, as it probably will, estimate the additional increment as closely as you can. For example, in Figure 2.22 the pointer is resting between the two minor divisions that represent 2.3 and 2.4. You can estimate the value of the unmarked increment by mentally dividing the space into halves, and halves of halves, as shown in the inset. Since the pointer is slightly to the right of the first fourth of the minor subdivision, you would estimate this reading to be 2.33.

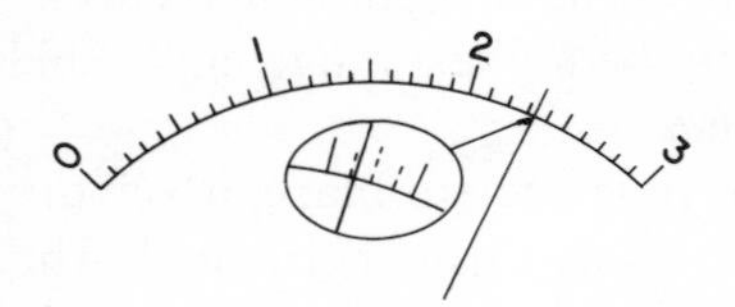

Figure 2.22 Reading a Meter Scale

After taking the reading, close the shorting switch, and if no more readings are to be taken, turn off the power, disconnect the meter, and reconnect the circuit where it was opened.

Voltage Measurements

Much of what was said about current measurements also applies to voltage measurements. Your purpose now is to obtain either (a) the total potential from a point in the operating circuit to ground; or (b) the potential *drop* between two points in an operating circuit.

For this you will require a DC or AC voltmeter as applicable, with a high-enough ohms-per-volt rating so it will not load the circuit unduly, and with a range appropriate to the anticipated voltage. Your preparations will be the same as for measuring current, except you do *not* open the circuit, and do *not* connect a shorting switch.

If it is not possible to turn off the power before connecting the meter, be sure and connect the ground (common) terminal first, and disconnect it last. Use an insulated test prod for the other terminal, and avoid physical contact with the metallic part. For measuring the potential drop between two points where neither is at ground potential, use two prods, and remember the whole meter may have voltage on it.

For high-voltage measurements, use a high-voltage probe and extreme care. (High voltage is any potential over 500V; any potential over 50V may be dangerous, however.)

SUMMARY

In this chapter, you have covered the subject of direct-reading devices, of which the most important are the analog meters. The most generally used of these is the PMMC-movement type, often called the d'Arsonval meter. Though actually a current meter, the PMMC movement can also be used as a voltmeter, both AC and DC, and as an ohmmeter. You also read about iron-vane, electrodynamometer, hot-wire and electrostatic movements. All the important meter characteristics were defined. And finally, you read about how to use current and voltage meters.

3

TROUBLESHOOTING BY MEASURING RESISTANCE

The property of opposing the passage of an electric current, causing electric energy to be transformed into heat, is shared in varying degrees by all electronic components, but specifically by resistors.

Resistors, which outnumber all other components in electronic circuits, are made of *composition* (graphite and clay, often called "carbon"), *resistance wire* (alloy of copper, iron, nickel, chromium, zinc or manganese), *thin film* (carbon or metal deposited on ceramic), or *conductive plastic*. They may be fixed or variable.

Apart from their physical makeup, resistors are described by their resistance values (ohms), ability to withstand heat (wattage), and tolerances. Most resistors have fairly wide tolerances (5%, 10%, 20%). Those whose values are held to closer limits are called precision resistors, and are usually wire-wound or metalized.

Variable resistors are known as potentiometers ("pots") for adjusting voltage, and rheostats for adjusting current.

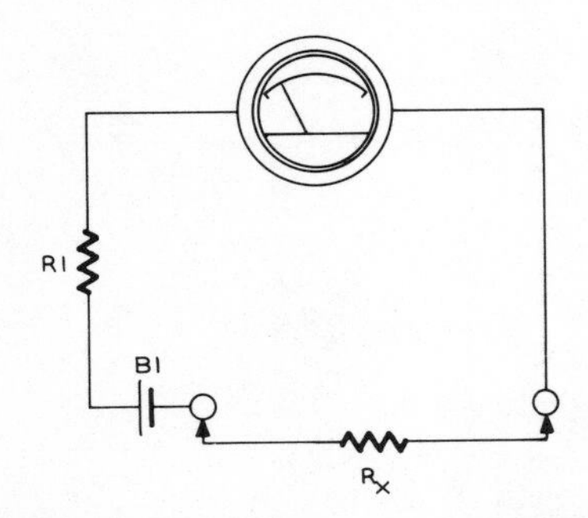

Figure 3.1 Principle of Series Ohmmeter

In addition to the measurement of resistors, resistance measurement is also used for conductors, transformers, capacitors, fuses, thickness of metals and metal coatings, cryogenic temperatures, breakdown resistance of insulation, and many other applications. This is a very broad range, with values of resistance from fractions of microhms to thousands of terohms.

To measure resistance we must furnish a current, since a resistor is a passive device. We can then measure the current flowing through the resistor, or the potential drop across it. An *ohmmeter* uses the first method; a *resistance bridge* the second.

OHMMETERS

An ohmmeter is more rapid in use but less accurate than a bridge. In essence, it consists of a reference voltage in series with a current meter and the resistance to be measured, as in Figure 3.1. R_X is the "unknown" resistance, connected to the ohmmeter's input terminals, and R_1 is a current-limiting resistor to prevent the current through the moving coil of the meter movement exceeding the value for full deflection when the input terminals are shorted together.

When this happens, the meter should read "O ohms." The zero end of the scale is, therefore, at the right-hand end, as shown in Figure 3.2. When nothing is connected to the input terminals, no current flows and the pointer stays at the left-hand of the scale, indicating "∞ ohms." When a resistance is connected to the input, the meter pointer deflects in proportion to the current, which flows according to the total resistance in the circuit: $R_T = R_X + (M + R_1)$. If the combined resistance of the meter and R_1 is 20 kilohms, and the reference voltage is a constant 1 volt, the current with the input shorted is 50 microamperes, which gives a full 90° deflection. Increasing values of resistance connected to the input reduce the current, and consequently the deflection, as shown in the following table:

R_X	$(M + R_1)$	R_T	$I (E/R_T)$	Deflection
0 Ω	20kΩ	20kΩ	50 μA	90°
20kΩ	20kΩ	40kΩ	25 μA	45°
40kΩ	20kΩ	60kΩ	16.7μA	30°
60kΩ	20kΩ	80kΩ	12.5μA	23°
80kΩ	20kΩ	100kΩ	10 μA	18°
100kΩ	20kΩ	120kΩ	8.3μA	15°

This table makes it clear why the scale illustrated in Figure 3.2 is so compressed toward the left-hand end. From 0 to 20 kilohms the pointer travels 45°, but from 80 to 100 kilohms only 3°; and from there to "infinity" only 15° more. Consequently, only the right-hand part of the scale is really useful for measurement.

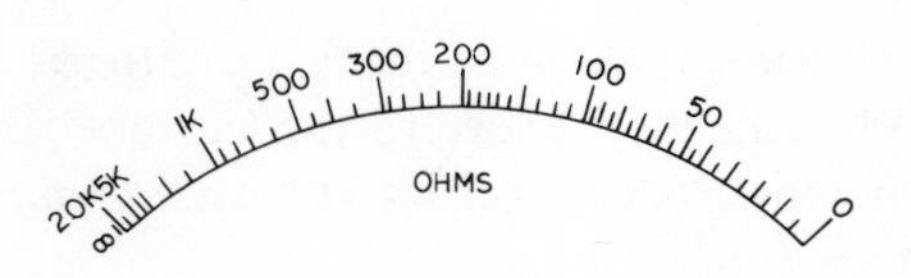

Figure 3.2 Series Ohmmeter Scale

The accuracy of this ohmmeter depends upon the quality of the meter and condition of the battery. The meter's characteristics are constant, but the battery ages, and its voltage changes with changes in the load. To compensate for this a variable resistor R2 is connected parallel to the meter, as shown in Figure 3.3, so that some of the battery current may be bypassed around it. If the battery potential is increased to 1.5V, we can adjust R2 so that when the input terminals are shorted together the meter current will be 50μA as before, with the balance going through R2. Before taking a reading with this ohmmeter, you must short the input terminals or test leads together, and adjust the "ohms adjust" control so that the pointer will rest on the zero mark of the scale. If the pointer will not zero with maximum rotation of the ohms control, the battery should be replaced.

Ohmmeters usually have several ranges so that measurements of higher and lower resistance values may be made on the right-hand part of

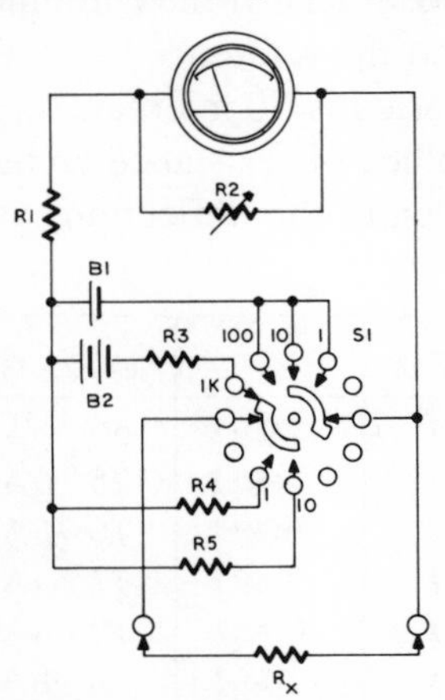

Figure 3.3 Complete Series Ohmmeter Circuit

the scale, where the best resolution is. Figure 3.1 is actually the X100 range circuit of the ohmmeter in Figure 3.3. For resistance up to 20kΩ measured on this range the pointer will deflect in the right-hand segment. However, for a resistance of only 200 ohms the pointer would indicate only slightly to the left of the zero mark. By rotating the range switch S1 to the X10 position, resistor R5 bypasses a portion of the meter current around the meter, so that the pointer deflects somewhat less (to the second higher mark to the left of zero on the scale in Figure 3.2). When the range switch is rotated to the X1 position, resistor R4 bypasses a more considerable portion of the meter current, so the pointer now deflects only to the 200-ohms mark. Obviously, this range will give the best resolution for low values of resistance.

On the other hand, higher resistances also have to be moved as much to the right as possible. This can only be done by increasing the current through the meter, so B2, which has a potential of 15V, is substituted for B1. R3 has a value that, added to the resistance of the meter and R1, makes R_T 200kΩ, which limits the current as before to 50μA. When a resistance of 200kΩ is now connected to the input terminals, the pointer deflects halfway (to the 200 mark) because the internal and external resistances are equal.

The ohmmeter just described is a *series-type* ohmmeter, because the unknown resistance is connected in series with the meter. In a *shunt-type* ohmmeter, R_x is connected in parallel with the meter as in Figure 3.4. As you can see, in this circuit maximum current flows through the meter when nothing is connected to the input, and no current flows through it when the input is shorted. The meter scale is therefore reversed from that in the series type, with the better resolution now at the left-hand end. Deflection is also greater for lower resistances. However, this ohmmeter is not used in direct-reading instruments of the kind found in portable

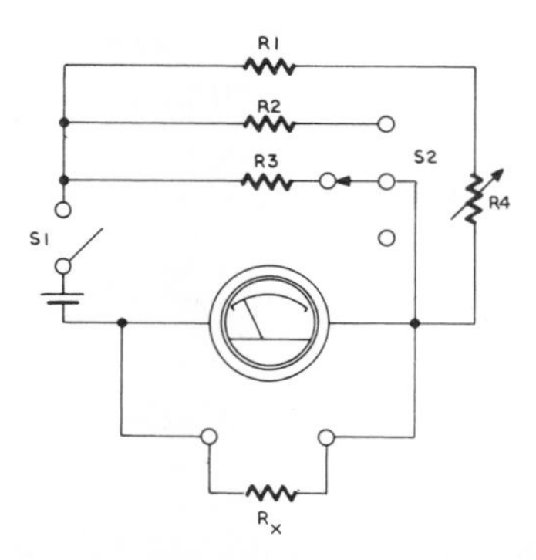

Figure 3.4 Shunt-Type Ohmmeter

battery-powered multimeters, because the circuit is heavy on the battery, which is too easily run down by leaving the meter switched on.

MEGOHMMETERS

High values of resistance cannot be measured very well by an ohmmeter because of the limited reference voltage available. To obtain full-scale deflection when reading hundreds of megohms requires thousands of volts. For this reason, special instruments called *megohmmeters* or "meggers" are used for measuring the resistance of insulation, earth grounds or other ultra-high resistances.

Standard megohmmeters are capable of generating up to 1,000 volts, and can measure resistance values up to 2,000 megohms. To generate the voltage a hand generator is built into the instrument. The meter has a special type of crossed-coil movement, as shown in Figure 3.5. The *deflecting coil* is in series with the generator, and the unknown resistance. The *control coil* is in series with the generator only.

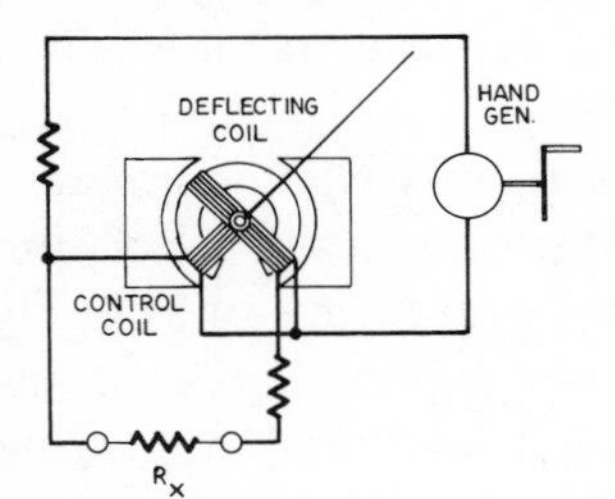

Figure 3.5 Megohmmeter with Hand Generator

With the input terminals shorted together, the current through each coil would be the same, but as their fields oppose each other the pointer does not move. When a resistance is connected across the input, the current through the deflecting coil changes, and the pointer now moves in proportion. The advantage of this arrangement is that the actual voltage of the generator output does not matter, since the meter is affected only by the change caused by the resistance.

There are several other instruments used for the measurement of high resistance values. These do not use hand generators, but regulated electronic power supplies capable of an output of many thousands of volts.

VOLT-AMMETER METHOD OF MEASURING RESISTANCE

The resistance value of a resistor can be found by measuring the current flowing through it and the voltage dropped across it, as shown by

Figure 3.6. The voltage read on the voltmeter, divided by the value of the current read on the ammeter, gives the resistance. However, for proper accuracy you should allow for the resistances of the meters, using the applicable formula, where:

E = voltmeter reading
I = ammeter reading
RA = internal resistance of ammeter
RV = internal resistance of voltmeter

The accuracy of resistance measurements made by this method is equal to the sum of the individual accuracies of the meters used.

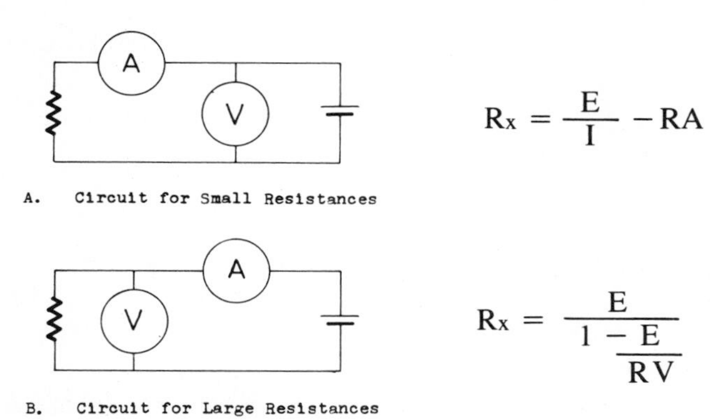

Figure 3.6 Volt-Ammeter Method of Measuring Resistance.

RESISTANCE BRIDGES

If each of the four resistors in the *Wheatstone bridge* diagrammed in Figure 3.7 is exactly 100 ohms, the voltage at X must equal that at Y. If A and B are 100 ohms each, while R_s and R_x are both 200 ohms, the voltages at X and Y must also be equal. And if A is 200 ohms while B is 100 ohms, the voltages at X and Y will still be equal as long as R_x has twice the resistance of R_s. Mind you, this does not mean that the voltage values will stay the same as the ratios are changed, but as long as $\frac{A}{B} = \frac{R_x}{R_s}$ the value at X must equal the value at Y.

Since you are primarily interested in measuring the "unknown" resistance R_x this equation can be expressed more conveniently in the form:

$$R_x = \frac{R_s\,A}{B}$$

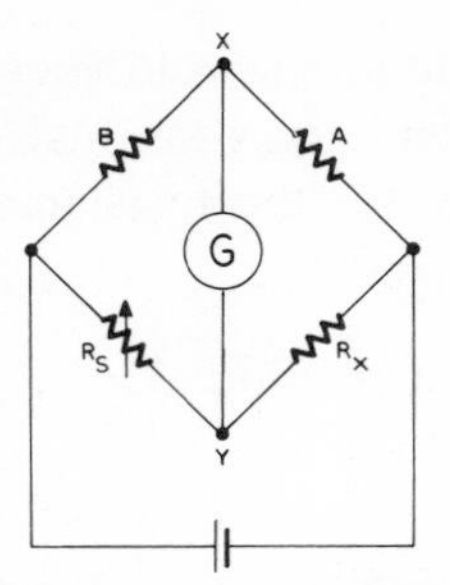

Figure 3.7 Wheatstone Bridge

Then, if you know R_s and the ratio A/B, you can calculate R_x.

However, this equation applies only when the potentials at X and Y are equal. Consequently, R_s is variable, so that it can be adjusted until its resistance, multiplied by the ratio A/B, equals R_x.

When the X and Y potentials are equal, the galvanometer G indicates zero current between them. This is called a *null*, hence the galvanometer is frequently called a *null detector*. A null detector may be built in as part of a bridge instrument, or may be a separate meter with a moving-coil, spotlight movement as described in Chapter 2. An electronic null meter with an amplifier gives even greater sensitivity. DC null detectors have meter scales with zero in the center, and the null condition is indicated when the pointer or light-spot is exactly on zero.

The variable resistor R_s is called the *rheostate arm* or *variable standard* of the bridge. In many bridges, it consists of from four to six *precision resistance decades*. These are instrument-quality rotary switches, on each of which are mounted ten precision wirewound resistors. The dials on these switches have positions numbered from 0 through 10 (to allow overlap of the ranges for greater convenience). Four dials could let you select resistances up to 11,110 ohms in 1-ohm steps; or five dials might allow selections up to 11,111.0 ohms in 0.1-ohm steps; and so on, the actual totals and increments varying with the model of the bridge.

The A and B arms are called the *ratio arms* or *multiplier* of the bridge. These resistors also may be selected by rotary switches, but in some bridges they are selected by *plug switches*, in which brass pegs are inserted firmly between brass blocks to make the desired connections. Ratios may be selected from 1/10000 through 1/1 to 10000/1, in multiples of 10 or otherwise, depending upon the model of the bridge.

The Wheatstone bridge is most sensitive when the A/B ratio is 1/1, so you should use this ratio whenever you can. However, if R_x has a value outside the range of the rheostat arm, you will have to use the multiplier. This may require an increase in the battery voltage from a minimum of 1.5V up to as much as 45V, depending upon the resistance being measured. However, you must find out what the specified safe current for the bridge is, *and be careful not to exceed it*. Many bridges have an internal battery of 4.5V or thereabouts for general use, and input terminals for an external battery when required. This insensitivity at higher ranges limits the Wheatstone bridge to a maximum resistance measurement in the vicinity of 10 megohms, although theoretically it could go ten times higher. The lower limit is about 0.1 ohm, but the limiting factor in this case is the contact resistance of the switches, or *switch zero resistance*. The total switch zero resistance of a 6-dial bridge may be of the order of 0.0035 to 0.004 ohm, which can have a considerable effect on resistance measurements lower than 0.1 ohm.

Even if the switch zero resistance is negligible, there is another difficulty. Suppose you want to measure a resistance of 0.001 ohm, and you require the value to four places. The bridge has a four-dial rheostat, the highest being X1000. The ratio between R_s and R_x is, therefore, 1000 : 0.001, or $10^6 : 1$. To balance this the A/B ratio will have to be $10^6 : 1$ also. You can see that almost the entire battery voltage will be dropped across R_s and B.

The 1000-ohm resistor in the rheostat arm, with a wattage of 0.5W, cannot take more than 22.36mA, which restricts the battery voltage to 22.36V. Since the voltage at Y will be approximately one millionth of this, it will give no indication at all on the galvanometer.

By replacing the rheostat-arm with a slide-wire, or bar of manganin with a sliding contact, the ratio between the standard and unknown resistances can be brought into better proportion. However, the current increase makes the voltage dropped across the resistance of the sliding contact larger, so the gain from the increased sensitivity is counterbalanced by the increased switch zero resistance.

The *Kelvin bridge* gets around this difficulty by adding two additional arms, *a* and *b* as shown in Figure 3.8. As long as the ratio a/b equals A/B, R_x still equals R_sA, but because the resistance values of a and b are so much higher than R_s and R_x practically all the current flows

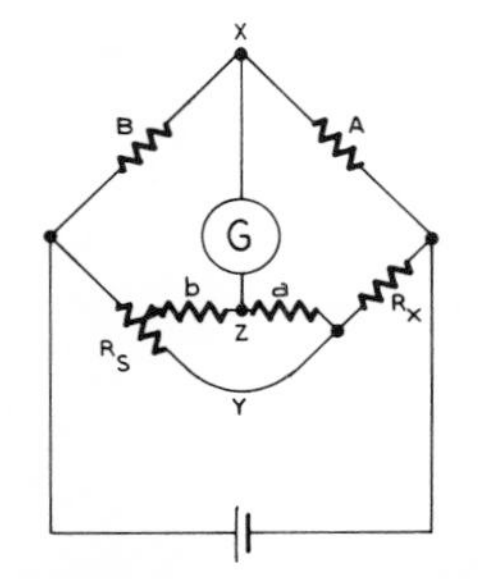

Figure 3.8 Kelvin Bridge

through R_s – Y – R_x. That through b – Z – a is negligible, and therefore, the voltage dropped across the sliding contact is no longer a problem.

The Wheatstone and Kelvin bridges between them are able to handle resistance measurements from .001 ohm to over 10 megohms. Because of their similarity, they can be combined in one instrument, as shown in Figure 3.9.

Figure 3.9 Kelvin-Wheatstone Bridge *(Courtesy Shallcross)*

This is a compact, portable bridge for general use. It has a wooden case with detachable lid. The panel is of heavy phenolic, and the rugged rotary switches have silver contacts and steatite insulation. The resistors are of precision wirewound construction that changes very little with temperature, and are non-inductively wound. There is a built-in galvanometer and battery.

USING THE KELVIN-WHEATSTONE BRIDGE

To measure the resistance of a resistor on this bridge is simple. Suppose you have one of approximately 5 kilohms, and wish to know its value with much greater accuracy than is possible on an ohmmeter.

It saves time if you know the approximate resistance before you start. You can believe the value marked on the resistor if you wish, or alternatively, you can make a quick preliminary measurement with an ohmmeter. Let's assume the ohmmeter indicates just under 5 kilohms.

Place the bridge on the bench in a horizontal position, and open and remove the lid. Check the galvanometer zero by sliding the slide-lock button away from the knurled knob. The pointer should rest exactly over the center of the scale. If it does not, loosen the screws on each side of the knurled knob, adjust the knob until the pointer is over the centerline, and then retighten the screws.

Check if the INT BA/EXT BA battery switch is in the internal (INT BA) position, and that the K-W switch is in the W (Wheatstone) position. Depress the BA (battery) button, and check if the galvanometer pointer deflects fully. If it does not, the internal battery should be replaced.

Connect short leads to the WHEATSTONE X terminals, and to the resistor to be measured. Lead resistance can be ignored for a resistance of this magnitude. Rotate the RATIO knob to the W 1 position (this makes the Wheatstone A/B ratio 1/1). Rotate the 1000 rheostat knob to the 4 position, the 100 knob to the 9 position, and leave the 10 and 1 knobs at 0.

Depress the LOW SENSITIVITY push-button momentarily. The meter pointer will deflect + or −. If +, turn the 100 knob to 8, and depress the button again. If it now deflects −, turn the 10 knob to 1 and push the button again. It should deflect less this time. Turn the knob to 2, and repeat. Continue until the pointer is at or near zero. If the initial deflection is −, you should leave the 100 knob set to 9, and begin to adjust the 10 knob from 1 on up as required. If you reach 10 without nulling the meter, return the 10 knob to 0, and readjust the 1000 knob to 5 and the 100 knob to 0. Then start adjusting the 10 knob again.

When the pointer is at or near zero when depressing the LO SENSITIVITY push-button, change to the HIGH SENSITIVITY button. Your adjustments will now be on the 1 knob. When the meter is nulled, or as close to a null as you can get it, read the resistance value from the positions of the four rheostat dials. If, for example, they are:

1000	100	10	1
4	9	5	6

the resistance value is 4956 ohms. Had the RATIO knob been set at some position other than W1, it would have been necessary to multiply this reading by the ratio indicated to obtain the value.

For values of resistance that are in the upper or lower portions of the usable range, the procedure is similar, except that because the bridge is much less sensitive an external battery should be used. External batteries may be up to 45V, depending on the resistance value being measured. No external battery should be used without a current-limiting resistor in series with it. A rule-of-thumb for the value of the resistor is $R = \frac{E^2}{4}$, where E is the battery voltage. For a 45V battery, the resistance would be about 500 ohms. The terminals for external power are on the left-hand side of the panel. When using an external battery, the INT BA/EXT BA switch must be set to EXT BA.

The values read on the rheostat dials must be multiplied by the setting of the RATIO knob, as already mentioned. With the knob set at W 1000, the rheostat dial setting of 4956 in the previous example will give a resistance value of 4956 x 1000 = 4.956 megohms. At the other end of the range, if the RATIO knob is set on W.001, the same rheostat-arm reading will be 4956 x .001 = 4.956 ohms.

When you come to measuring low resistances on the Kelvin-bridge part of the instrument, you must take into account the lead and connection resistances that you were able to ignore when measuring higher values on the Wheatstone bridge. These stray resistances change their values considerably with even small changes in room temperature. Connection resistance also varies with the force applied to the terminals and with the amount of oxidation or dust present. Lead resistance can be allowed for by taking a preliminary reading with the leads shorted together. This value is then subtracted from subsequent measurements. However, this is not necessary in the *four-terminal method* we are about to discuss.

Just now you saw how Lord Kelvin solved the contact-resistance problem by routing most of the battery current through R_s – Y – R_x, balancing the bridge by equalizing the voltages at X and Z, Z being in the path b – Z – a with negligible current. The same principle is used in connecting the resistance to be measured to the bridge. For Kelvin measurements, there are four terminals. The two marked C are current terminals. Those marked P are the potential or voltage terminals. A resistor being measured is connected as in Figure 3.10, in which the actual connections to the resistor leads are made with *Kelvin clips*. These are spring

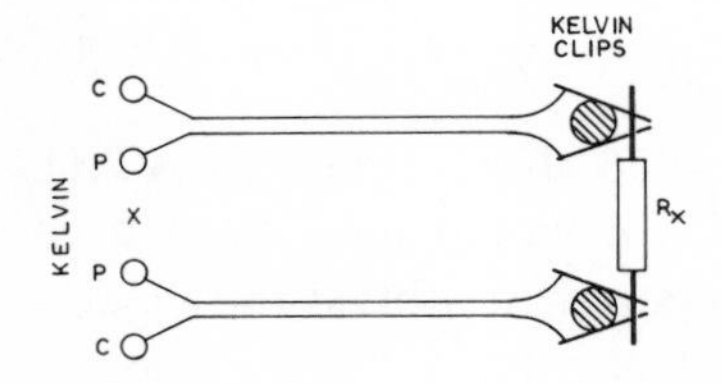

Figure 3.10 Four-Terminal Resistance Measurement Using Kelvin Clips

clips with jaws insulated from each other. In each clip, one jaw is the current connection, the other the potential. Since the important potential drops due to stray resistances are all in the current path, while those in the potential path are negligible because the current is almost nonexistent, these clips allow you to make accurate measurements even when the resistance being measured is lower than the resistance of the clips.

Kelvin clips enable you to make precision resistance measurements rapidly, but four ordinary leads with alligator clips work just as well, as long as they are connected in the same way.

Because of the low values of resistance you need a battery with high-current capability for Kelvin measurements. You should use a 2V storage battery, therefore, instead of dry cells. A rheostat for limiting the current is preferable to a fixed resistor, as it allows you to make adjustments without disconnecting it from the circuit.

The majority of the panel controls are used in the same way as for Wheatstone measurements, except that the K-W switch is changed to K, and the K ratios are used in setting the RATIO control.

The built-in galvanometer has a maximum sensitivity of μA per mm of scale deflection. However, you can get greater sensitivity with a moving-coil, spotlight type galvanometer, as described in Chapter 2, with a sensitivity of 0.005μA or better. You connect this to the GA terminals, of course. The built-in nullmeter can be used first to obtain a rough null.

The characteristics of the galvanometer you select will make a lot of difference as to how well it will work for the measurement you are making. Its sensitivity should be just sufficient for the purpose. Excessive sensitivity makes it too lively, and you waste a lot of time waiting for it to settle down each time you depress the sensitivity push-button. On the other hand, insufficient sensitivity prevents you from getting all the accuracy you want. Response time, or period taken to settle down after each adjustment, should be as short as possible, consistent with other desirable characteristics.

Damping should be slightly under- rather than over-damped. This is the case where the internal resistance of the galvanometer movement and its external critical damping resistance add up to about three-fourths of the circuit resistance as "seen" by the galvanometer. (The meaning of these characteristics was explained in Chapter 2.)

ELECTRONIC NULLMETERS

A good galvanometer has a sensitivity in the low nanoampere range, but an electronic nullmeter can extend this down into the low picoampere region, so that it is a thousand times more sensitive. The modern instrument is typically a battery-operated, solid-state zero center-scale electronic voltmeter, with a superior PMMC movement. We shall discuss this type of instrument in more detail in Chapter 6.

OTHER RESISTANCE-MEASURING INSTRUMENTS

The difficulties encountered at the upper and lower ends of the resistance spectrum have led to the development of special instruments for use in these areas. If you have a continuous requirement for making such measurements, such an instrument will be faster and more convenient in use, and does not require as much skill in the operator.

For very high resistance measurements, where the megohmmeter would not be accurate enough, you can use a *megohm bridge,* or a *conductance bridge*. A megohm bridge is a Wheatstone bridge designed for high resistance measurements, and its principle of operation is the same. A conductance bridge is also a Wheatstone bridge, but rearranged to measure *mhos* instead of ohms, since conductance is the reciprocal of resistance.

As you can see in Figure 3.11, in the conductance bridge ratio arm B changes place with the rheostat arm R_s. The bridge equation now becomes $\frac{A}{Rx} = \frac{R_s}{B}$, so that $\frac{1}{Rx} = \frac{R_s}{AB}$. The term $\frac{1}{Rx}$ is G_x, of course, the conductance of the unknown resistance.

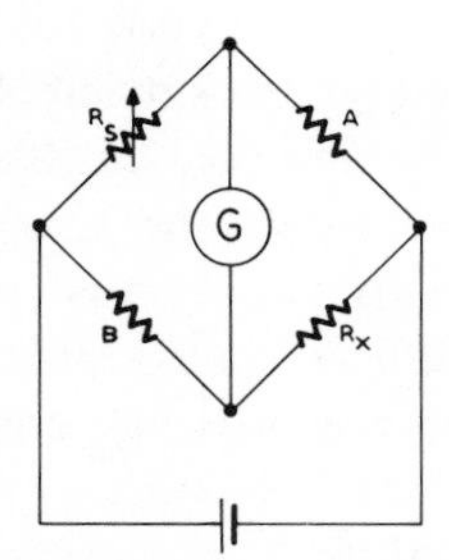

Figure 3.11 Conductance Bridge

A conductance bridge can measure resistance up to 10^{-15} ohms (10^{-15} ohms), provided it is used with a suitable null detector, and the right bridge voltage. For measurements that approach the limits of the "state of the art" of resistance measurements, a strip-chart recorder should be used in conjunction with the null detector, as the recorded value over a period of time doubles the resolution the human eyeball is capable of.

THREE-TERMINAL HIGH-RESISTANCE MEASUREMENTS

When you connect a resistor between two terminals mounted on a terminal board of insulating material, the *leakage resistance* of the board shunts the resistor just as if were also a resistor. Its resistance is typically in the terohm range, so for ordinary levels of resistance measurement it is of no importance. But in very high-resistance measurements it cannot be ignored.

This shunting resistance can be eliminated by connecting the resistor being measured to two insulated terminals mounted on a metal plate, instead of on insulating material. The metal plate is then connected to a third terminal on the megohmmeter or bridge designated the *guard* terminal. In the Wheatstone bridge in Figure 3.7, the guard terminal would be connected to X.

The leakage from R_x to the metal plate through the terminal nearest to A is now in parallel with A, while the leakage through the other terminal is in parallel with the galvanometer G, so neither shunts R_x, and therefore has no effect on its measured value. A bridge arranged so that the leakage resistance does not shunt the resistance being measured is said to be "guarded."

MILLIOHMMETERS AND LIMIT BRIDGES

At the lower end of the resistance spectrum, the *milliohmmeter* is the counterpart of the megohmmeter. Just as the megohmmeter is a streamlined Wheatstone bridge, so the milliohmmeter is a modified Kelvin bridge, indicating the resistance value on a panel meter. Neither instrument is as accurate as a standard bridge.

Manufacturers of resistors or other components whose resistance must meet close tolerances make use of *limit bridges* for rapid production testing. These are the same as standard Wheatstone bridges, but have two additional dials for setting the percentage limits (plus or minus) of the specified tolerance. The operator sets the nominal value of the resistors he is checking on the rheostat dials, the permissible deviation on the percent dials, and connects his resistors one by one. The direction of deflection of the null meter tells him whether to accept or reject each one.

SUMMARY

This chapter has covered the principal methods used for the measurement of resistance. We described the procedure for using a Kelvin-Wheatstone bridge, as this is the type of instrument most likely to be used for accurate resistance measurement. The use of the ohmmeter was also covered, and will be alluded to again when we come to multimeters.

The methods and instruments mentioned in this chapter are summarized in the following table.

TYPE	RANGE IN OHMS	TYPICAL ACCURACY %	USE
Kelvin Bridge	10^{-6} - 10^{1}	0.5	Laboratory
Milliohmmeter	10^{-5} - 10^{2}	3	Portable
Ohmmeter	1 - 10^{6}	3	Portable
Wheatstone Br.	1 - 10^{7}	0.1	Laboratory
Conductance Br.	1 - 10^{10}	0.05	Laboratory
Megohmmeter	10^{4} - 10^{14}	3	Portable
Megohm Bridge	10^{3} - 10^{15}	1	Laboratory

The portable instruments are faster and more convenient in use than the laboratory instruments, which require more care in order to take full advantage of their greater accuracy.

In the next chapter, we shall cover impedance bridges. Some of these are *universal*: that is to say, they also measure resistance in addition to inductance and capacitance. However, they generally use the Wheatstone bridge for this in the same way as you have seen in this chapter.

4

HOW TO MEASURE IMPEDANCE

In Chapter 1, we said that there is no such thing as a pure resistance, inductance or capacitance in actual practice. All resistors, inductors and capacitors have resistive, inductive and capacitive elements, of which the combined effect is impedance. Then, in Chapter 3, we went through the whole business of measuring resistance without ever once mentioning inductance or capacitance!

We were able to ignore the inductive and capacitive elements of resistance because we were using DC in our measuring instruments. This cannot be done when measuring inductance or capacitance, since AC must be used. Therefore, an impedance bridge has to balance both the reactive and resistive elements of the impedance it is measuring.

The components of impedance (Z), as shown in Figure 4.1, are:

R = resistance $\qquad$ θ = phase angle

X = reactance $\qquad$ δ = loss angle

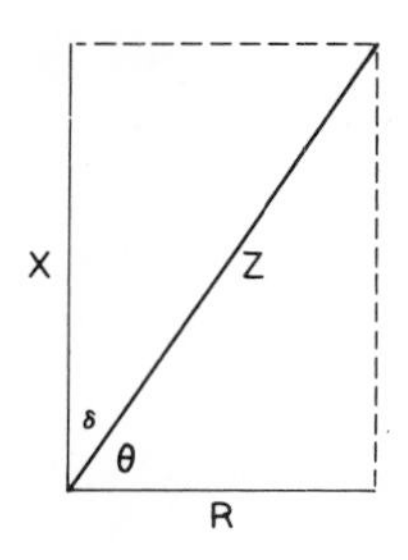

Figure 4.1 Components of Impedance

A good inductor will have a lot of X and very little R, which in turn will make θ large and δ small. This inductor will have an excellent *storage factor* (*Q*) because:

$$Q = \frac{X}{R} = \tan\theta = \cot\delta$$

The opposite of Q is the *dissipation factor* (*D*), which, of course, will be small:

$$D = \frac{R}{X} = \cot\theta = \tan\delta$$

The higher the resistive element R with respect to the total impedance Z, the greater the proportion of applied power that will be consumed in overcoming the resistive element. This ratio is the *power factor* (usually expressed as a percentage):

$$PF = \frac{R}{Z} = \cos\theta = \text{sine}\ \delta$$

These factors also apply to capacitors, except that X is negative and you are more interested in D than Q.

Reactance also varies with the frequency ($X_L = 2\pi fL$, $X_C = \frac{1}{2\pi fC}$). If you know the reactance and frequency, you can calculate the inductance or capacitance. Since all these factors are mathematically interrelated, you can see that impedance bridges can measure:

Inductance	L	Reactance	X
Capacitance	C	Phase Angle	θ
Resistance	R	Loss Angle	δ
Storage Factor	Q	Frequency	f
Dissipation Factor	D	Power Factor	PF

Universal impedance bridges are designed to measure the first five factors. The remainder may then be calculated. Other instruments are also available for measurement of individual factors with greater precision. Not all of them are impedance-measuring devices; for example, frequency is more often measured with an electronic counter.

UNIVERSAL IMPEDANCE BRIDGE

Figure 4.2 illustrates a typical universal impedance bridge, and Figure 4.3 shows its block diagram. You'll notice at once that the connections of the detector and generator are opposite to those of the galvanometer and battery shown in Figure 3.7. This does not matter at all. The Wheatstone bridge works just as well both ways.

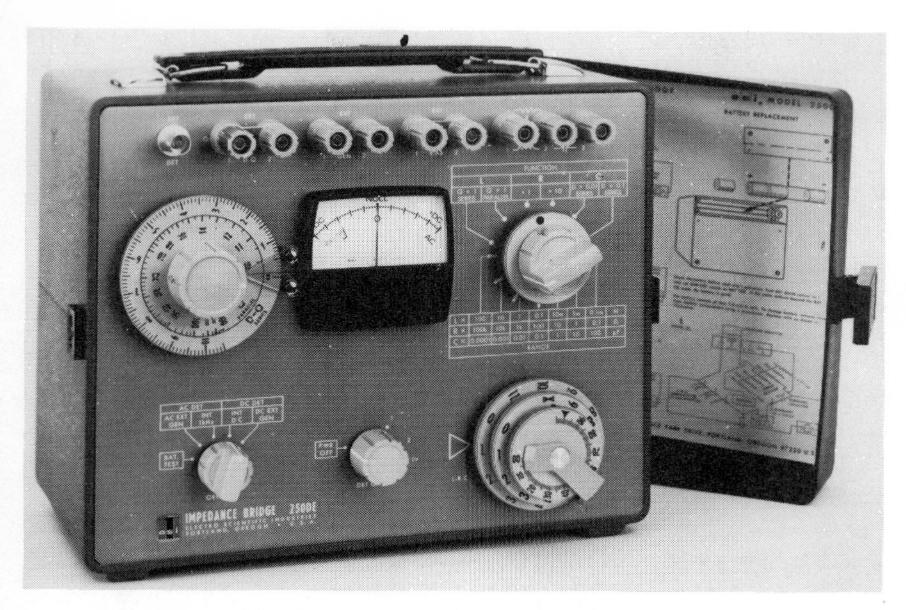

Figure 4.2 Universal Impedance Bridge *(Courtesy ESI)*

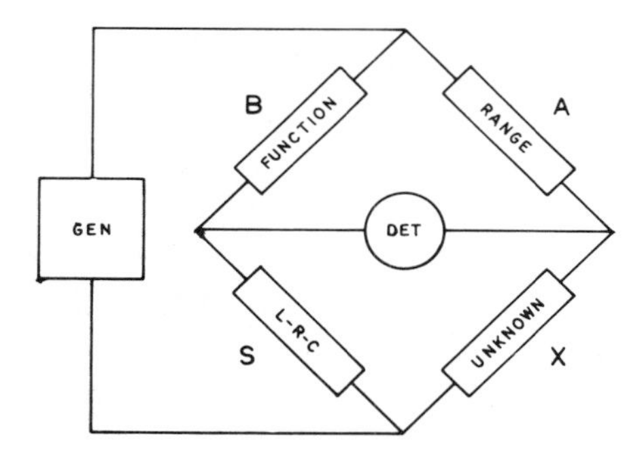

Figure 4.3 Block Diagram of Universal Impedance Bridge

The "unknown" arm of the bridge is whatever is connected to the binding posts in the top right-hand corner of the panel (Figure 4.2). These terminals are marked to show which to use for resistance, inductance or capacitance. You'll see what the other terminals are used for as we go along, but notice that two pairs (EXT BIAS and EXT D-Q) are normally connected by shorting lugs.

The rheostat arm is the L-R-C control. This consists of a three-tier decade rheostat operated by three concentric dials. The outermost dial has 11 steps of 1 kilohm each, the middle dial has ten of 100 ohms each, and the innermost dial rotates a 105-ohm rheostat and has 105 minor divisions (one per ohm). The setting of the rheostat arm is read from the dial figures lined up opposite the arrow. This indicates the value of the resistance, inductance or capacitance after the bridge is balanced.

The rheostat-arm (L-R-C) reading is multiplied by the setting of the RANGE control, which selects the right value to establish the proper ratio. The FUNCTION control selects the elements for the fourth arm that will adapt the bridge for each type of measurement. When L or C functions are selected, the D-Q dial is connected into the circuit to help balance the bridge. Its reading after the bridge is balanced indicates the value of D or Q.

The bridge is battery-operated, and the AC DET/DC DET switch connects the battery or an external source for AC or DC operation. When in the AC DET/INT 1 kHz position an internal generator produces a 1-kilohertz signal for the bridge current. The BAT TEST position connects the null meter to the battery. The pointer should read in the BAT OK segment of the scale, and the battery must be replaced if it does not.

The DET/GAIN control operates in the same way as the on-off-volume control on a radio or television set. It turns the power on and adjusts the sensitivity of the detector.

DC RESISTANCE MEASUREMENT

First of all, turn the DET GAIN control to *2*, and the GEN DET control to INT DC. Then connect the resistor to be measured to the R and L terminals, making sure you have a good tight contact.

The FUNCTION switch connects a resistor into the B arm of the bridge when it is set to R. In the X1 position, this is a 10-kilohm resistor, in the X10 it is a 1-kilohm one, so when the FUNCTION switch is in the X10 position the RANGE control values are all ten times greater. The X10 position is required only for resistance measurements of over a megohm. The bridge circuit for resistance measurement is as shown in Figure 4.4, which you'll recognize as the familiar Wheatstone circuit.

If you already know the approximate value of the resistance you are measuring, you can set the RANGE switch to the appropriate position. For example, for a 5-kilohm resistor you'd set it to R X 1kΩ, and then adjust the L-R-C dials for a null on the meter, as explained below. If you don't know the value, first set the L-R-C dials to 3.000 and then rotate the RANGE switch until you find the position which gives the least deflection of the meter pointer. (If this is in the R X 100kΩ position, switch the FUNCTION control to R X 10 to see if this gives a smaller deflection. If it does, leave it on R X 10; otherwise return to R X 1.) Then adjust the L-R-C dials for a null.

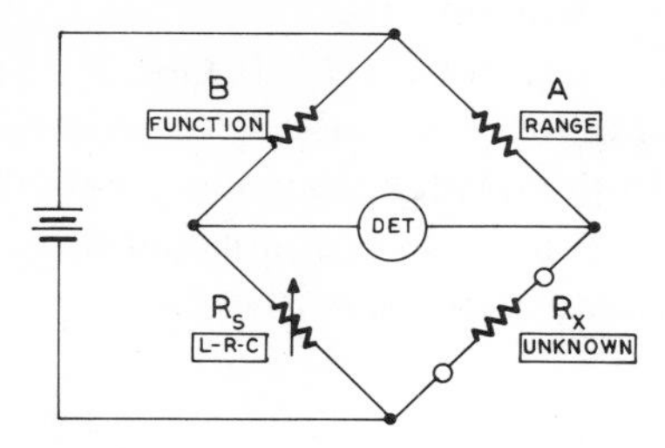

Figure 4.4 Universal Impedance Bridge Circuit for DC Resistance Measurement

In adjusting the L-R-C dials, rotate the outside one first until you find the position that gives the least deflection of the meter pointer. Then repeat with the middle dial and finish on the inside dial. If you get all the way to the upper or lower end of any dial without reaching a minimum on the meter, reset the next higher one up or down one step as appropriate, and try again. However, if this happens on the outside dial reset the RANGE switch. This should be done for any reading lower than 1.200 because the accuracy of the rheostat arm falls off below this value.

For very low values of resistance (RANGE R X 0.1Ω), if you are using leads to connect the resistor to the input terminals, you should first short the leads together and measure the lead resistance before connecting them to the resistor. This value is then deducted from the resistor's measured value to get its true value.

With high resistance values, make sure nothing that can affect the reading is shunting the resistor. Some insulation that is good enough for most purposes may provide considerable leakage when paralleling a 10-megohm resistor. If you decide to use an external battery, the same considerations apply as in the Chapter 3, when using the Wheatstone bridge described there. Set the GEN-DET control to DC EXT GEN and connect the battery to the EXT GEN terminals. *Don't forget the current-limiting resistor!*

CAPACITANCE MEASUREMENT

You may look upon the capacitive and resistive components of a capacitor as being in series (C_s) or in parallel (C_p). Since $C_p = \frac{C_s}{1 + D^2}$, there is little difference between them if D is less than 1. However, the bridge has to be designed to balance one or the other, and in this bridge they are regarded as being in series.

By connecting the unknown capacitor to the C input terminals, and setting the FUNCTION switch to C (D x 0.1 SERIES) we alter the basic Wheatstone bridge of Figure 4.4 to the series-capacitance ratio bridge shown in Figure 4.5. In this bridge, balancing is between the unknown arm and adjacent standard arm. When adjacent arms do the balancing, they are *alike*, so both are series-capacitive.

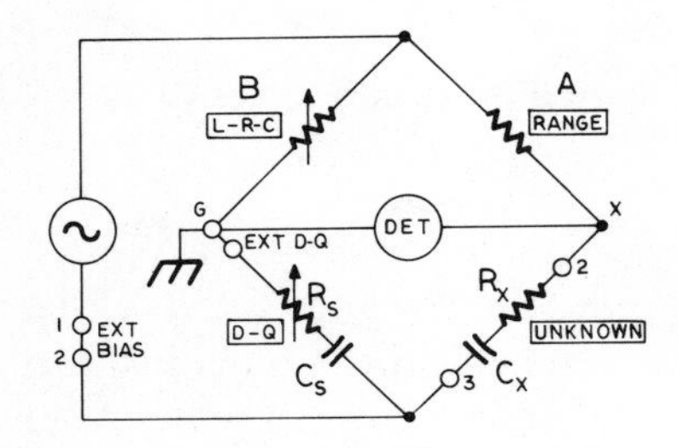

Figure 4.5 Universal Impedance Bridge Circuit for Series-Capacitance Measurement

The D-Q dial controls three rheostats, two of which are used for capacitance measurement. Setting the FUNCTION switch to C (D x 0.1 SERIES), selects the one preferred for a preliminary balance. (The other rheostat is substituted when the FUNCTION switch is turned to the D x 0.01 SERIES position.) Adjusting this rheostat, balances the R_x component of the "unknown" capacitor, and its reading, after balancing the bridge, gives the value of the capacitor's dissipation factor D.

C_s is a precision capacitor of 0.1 microfarad, and is the standard reference against which C_x is compared. It will balance C_x when the B/A ratio is such that the potentials at X and G are equal. The A resistance is selected by the RANGE control, and the B by the L-R-C dials. After balancing the bridge, the reading of the L-R-C dials, multiplied by the RANGE setting, gives the capacitance of C_x the microfarads.

In measuring the capacitance of a capacitor, you start by turning the DET GAIN control to *1*, which applies power to the instrument, and set the FUNCTION switch to C (D x 0.1 SERIES). Next, set the L-R-C dials to 3.000 and the D-Q dial to 0. Then connect the unknown capacitor to the C binding posts. Tighten the terminals for a good connection.

If it is not possible to attach the capacitor directly to the input terminals you should use a shielded cable to eliminate errors due to stray capacitance and EMI, as shown in Figure 4.6. Keep the leads as short as possible. The terminals 2 and G are shown in Figure 4.5 also, where you can see that G grounds the shield. The capacitance between the shield and the inner conductor is shunted across the detector, while any capacitance between the unshielded load and ground is shunted across C_s. As long as this does not exceed 10 picofarads, its effect on C_s is negligible, because the value of C_s is at least ten thousand times greater.

If your capacitor has a very low value you should also:

A. Keep your hands away from it once you have connected it, to reduce AC pickup and stray capacitance;

B. Subtract the zero capacitance of the bridge from the measured value. This is the dial reading (approximately 2 picofarads) with nothing connected to the input.

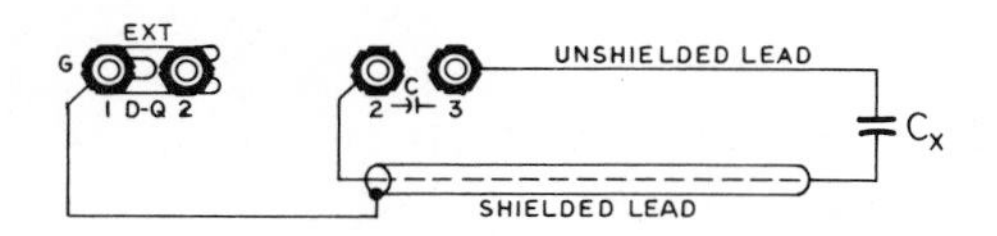

Figure 4.6 Reducing Stray Capacitance and EMI

After connecting C_x, set the GEN-DET switch to AC DET (INT 1kHz). This applies a 1-kilohertz signal to the bridge. (Other frequencies can be obtained by using an external generator, as explained later.) Rotate the RANGE control to find the position that gives the least deflection of the meter pointer, turning the DET GAIN control clockwise to increase sensitivity if necessary.

Now adjust the L-R-C dials for minimum deflection of the meter pointer, then do the same with the D-Q dial. Go back and readjust the L-R-C dials, then the D-Q; and continue adjusting the two dials alternately, increasing sensitivity with the DET GAIN control as required, until no further decrease in deflection can be got.

If the D-Q dial reads less than 1, reset the FUNCTION switch to C (D x 0.01 SERIES). Retouch the L-R-C and D-Q dial settings to renull the meter.

You can now read the capacitance of C_x from the L-R-C dials, multiplied by the RANGE switch position; and the dissipation factor from the D-Q dial reading multiplied by the FUNCTION switch setting. However, if the L-R-C dial reading is less than 1.200, move the RANGE switch one step to the left and repeat the nulling procedure, because you will get a more accurate reading.

If you have now finished, disconnect the capacitor and turn the DET GAIN control to PWR OFF.

SLIDING BALANCE

In this bridge, the reactive (C_x) and resistive (R_x) components of the unknown capacitor are balanced separately by the L-R-C and D-Q dials, so that:

$$C_x = C_s \frac{\text{RANGE}}{\text{L-R-C}}$$

$$R_x = \text{D-Q} \frac{\text{RANGE}}{\text{L-R-C}}$$

The standard capacitor C_s has a fixed value, so once you have set the RANGE control, the only adjustment required to balance C_x is that of the L-R-C dials.

Balancing R_x is then a matter of adjusting D-Q. When D is less than 1, D-Q is small compared to L-R-C. If L-R-C is readjusted to get a better null, D-Q will have to be readjusted also, but the change required will be small. The C_x null and the R_x null, which must coincide to give a true null, are never far apart.

But it's another story when D is large. A considerable rotation of the D-Q dial (or external rheostat, as explained later) is required to find the R_x null. When the L-R-C dials are adjusted for a minimum deflection of the null-meter pointer, the D-Q dial may be nowhere near the R_x null. Consequently, there will be a less definite null-meter indication as the L-R-C dials are adjusted. To make matters worse, the D-Q adjustment required to get an R_x null will change considerably as the L-R-C dials are adjusted. This is what is meant by the term "sliding balance." The frustrated technician, as he chases the elusive null, begins to feel that the bridge is playing games with him, even fooling him with false nulls at some dial settings!

You can outsmart the bridge, however. After getting a preliminary minimum indication (which may not be very sharp), adjust the DET GAIN control to place the pointer exactly over the third major-division mark to the right of zero. Now turn the outer L-R-C dial one step in either direction. If the meter indication increases, go back and turn one step in the other direction. If it decreases, adjust the D-Q dial for minimum deflection. Then readjust the DET GAIN control and repeat the process. This should bring you to the true null a step at a time. Each step should be shorter than the last when you get close to it.

At this time, it may not be possible to reset the DET GAIN control to the reference mark, so use a new one closer to zero. Also, change over to turning the D-Q dial part of a dial division at a time, readjusting the L-R-C dials for minimum deflection.

False nulls are most likely to appear at the ends of the range of the instrument, so be suspicious of any there.

Using the Extended D-Q Range

The D-Q dial goes up to a little over 10, which, multiplied by 0.1 if using the D x 0.1 range of the FUNCTION switch, only permits measurement of D to 1.05. If D is higher than this, the D-Q range may be extended by connecting an external rheostat is series with D-Q.

Use a precision resistance decade box, and connect it to the EXT D-Q terminals, after disconnecting the shorting lug. This makes it part of the standard arm of the bridge, as shown in Figures 4.2 and 4.5.

The bridge is balanced as described above, using both the external rheostat and the D-Q dial. Calculate D as follows:

D = 0.628 x resistance of external rheostat in kilohms + D-Q reading.

This formula will apply as long as you are using the internal 1-kHz generator. For any other frequency, you must multiply the above by the frequency in kilohertz.

MEASURING POLARIZED CAPACITORS

Many capacitors, such as electrolytic and tantalum capacitors, require a polarizing voltage, and may give erroneous readings without it. This is easily supplied from a battery connected to the EXT BIAS terminals, in the circuit shown in Figure 4.7.

A battery is preferable to an electronic power supply because it is completely isolated from ground. C_B is required to bypass the 1-kHz

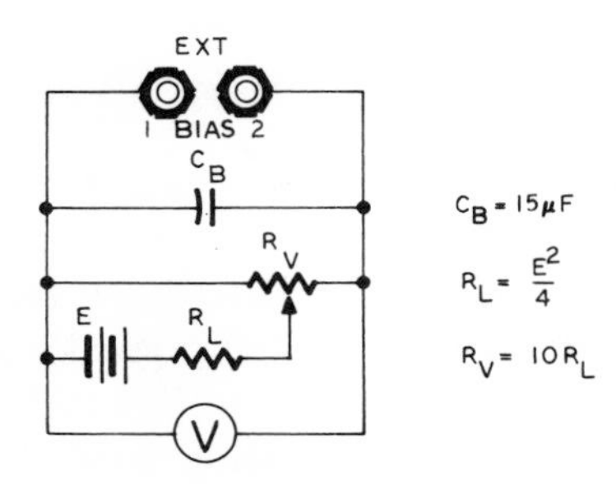

Figure 4.7 DC Bias Supply for Polarized Capacitors

signal around the battery, R_V is a potentiometer to adjust the voltage, and R_L is a current-limiting resistor to protect the bridge. You'll recall from the Chapter 3 that the value of the limiting resistor should be one-fourth of the square of the battery voltage. R_V should be approximately ten times R_L.

Connect the positive terminal of the DC bias supply to EXT BIAS TERMINAL 2 (disconnect the shorting lug), and connect the positive side of the capacitor to C input terminal 3.

INDUCTANCE MEASUREMENT

As in the case of capacitance, you may regard the inductive and resistive components of an inductor's impedance as being in parallel or series. However, as series inductance differs measurably from parallel inductance when Q is less than 100, the FUNCTION switch provides bridge circuits for both. The circuit for measuring series inductance is a Maxwell bridge, as shown in Figure 4.8, and for parallel inductance a Hay bridge, as shown in Figure 4.9. In these bridges, the balancing is done between opposite arms, consequently they are *opposite* in form: an unknown *series* inductance is balanced by a *parallel-capacitance* standard arm (Figure 4.8), and an unknown *parallel* inductance is balanced by a *series-capacitance* standard arm (Figure 4.9).

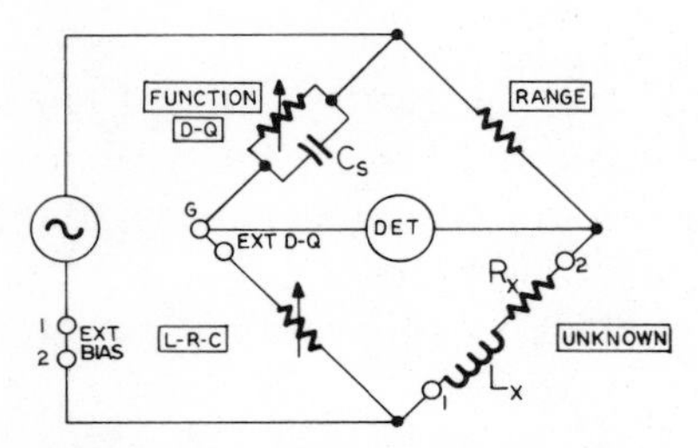

Figure 4.8 Universal Impedance Bridge Circuit for Series-Inductance Measurement

In measuring the inductance of an inductor, you start by turning the DET GAIN control to *1*, which applies power to the instrument, and set the FUNCTION switch to L (Q x 1 SERIES) if the inductor's Q is less than 10, or to L (Q x 1 PARALLEL) if Q is greater than 10. Next, set the L-R-C dials to 3.000, and the D-Q dial to 1000 or 10.5. Then connect the unknown inductor to the R or L binding posts (refer to Figure 4.2). Tighten the terminals for a good connection.

If it is not possible to attach the inductor directly to the input terminals, you should use short heavy leads, twisted together to minimize lead impedance. If you are making a low-inductance measurement, you should also take the same precautions as shown under capacitance measurement to eliminate AC pickup, stray capacitance and EMI.

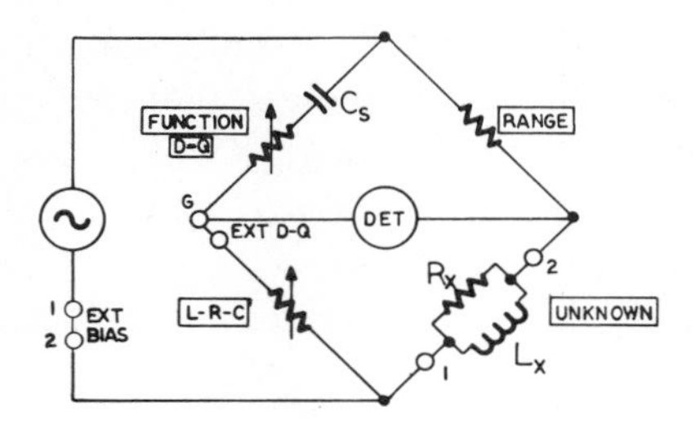

Figure 4.9 Universal Impedance Bridge Circuit for Parallel-Inductance Measurement

After connecting L_x set the GEN-DET switch to AC DET (INT 1kHz). Rotate the RANGE control to find the position that gives the least deflection of the meter pointer, turning the DET GAIN control clockwise to increase sensitivity, if necessary.

Now adjust the L-R-C dials for minimum deflection of the meter pointer, then do the same with the D-Q dial. Go back and adjust the L-R-C dials, then the D-Q; and continue adjusting the two dials alternately, increasing sensitivity with the DET GAIN control as required, until no further decrease in deflection can be got. If there is any difficulty in balancing the bridge, use the procedure explained earlier under *sliding balance*.

If the L-R-C dial reading is less than 1.200, move the RANGE switch one step to the right, and repeat the nulling procedure, because you will get a more accurate reading.

You can now read the inductance of L_x from the L-R-C dials, multiplied by the RANGE switch position; and its storage factor from the D-Q dial, using the inner dial for L-parallel and the outer for L-series.

If you have finished, disconnect the inductor and turn the DET GAIN control to PWR OFF.

Extended Q-Range Inductance Measurements

The range of Q that can be indicated on the D-Q dial when using the internal 1-kilohertz generator is from 0.1 to 1000. For values outside this range, an external rheostat must be used as in similar capacitance measurement. This is connected to the EXT D-Q terminals for parallel inductance, or between C input terminal and ground (EXT D-Q 1) for series inductance.

The unknown inductor is connected to C input terminals 2 and 3 for parallel inductance, L input terminals 1 and 2 for series inductance. Use the L (Q x 1 PARALLEL) position of the FUNCTION switch for all measurements, and turn the D-Q dial to 1000 and leave it there. (This shorts out the D-Q rheostat, so only the external rheostat is adjusted for balance.) The inductance, after balancing, will be the L-R-C reading multiplied by the RANGE setting.

For a parallel inductance, calculate Q from $Q = \frac{1}{2\pi fR}$, where f is the frequency in kilohertz (1, if using the internal generator) and R is the resistance in kilohms of the external rheostat. For series inductance the equation is $Q = 2\pi fR$.

Inductance Measurements with DC

Iron-core chokes and transformers are used frequently in circuits in which they pass DC as well as AC. If the DC is heavy enough to cause the core to approach saturation, the inductance decreases. Therefore, you might want to know the inductance of the iron-core inductor with DC also, rather than with AC alone.

Your DC supply should be set up as shown in Figure 4.10. Choose values of R_V and R_L as explained under capacitance measurement to avoid damage to the bridge. Connect to the EXT BIAS terminals (positive to terminal 2), and use the L-parallel setting of the FUNCTION switch, because with this setting the standard capacitor blocks the flow of DC through the bridge, so that the milliammeter measures the DC through the inductor.

External Generator

DC or AC supplies, connected to the EXT GEN terminals, may be substituted for the internal generator, if necessary. You must use a limit-

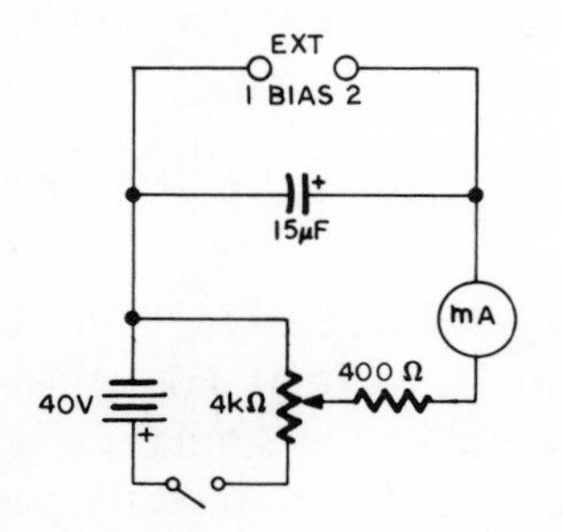

Figure 4.10 DC Bias Supply for Inductance Measurements

ing resistor just as for bias supplies. In addition, the AC supply voltage must not exceed one-twentieth of the frequency. Remember to set the GEN DET switch to the appropriate position.

When using an AC generator at a frequency other than 1 kilohertz, you must use an external detector tuned to the external frequency. Remember that the D or Q value found has to be multiplied by the frequency in kilohertz, except for parallel inductance, where you divide it. (The accuracy of the bridge may also be a little different at other frequencies. Be sure to check the manufacturer's manual for details.)

External Detector

An external detector may also be connected to the EXT DET coaxial connector, using a coaxial cable. This may be a more sensitive null detector or an oscilloscope. Since the internal detector is not disconnected, it may be used to obtain a preliminary null.

AC RESISTANCE MEASUREMENT

Although not as accurate as DC resistance measurement, it is sometimes important to find the AC resistance of a part. Proceed as for DC, except for setting the GEN-DET switch to INT 1kHz, but if you cannot get a sharp null, you will have to compensate for series inductance or parallel capacitance. As the DC Wheatstone bridge circuit makes no provision for this, an external capacitor and potentiometer must be connected as shown in Figure 4.11. You then adjust the potentiometer and the L-R-C dials alternately until you get a satisfactory null.

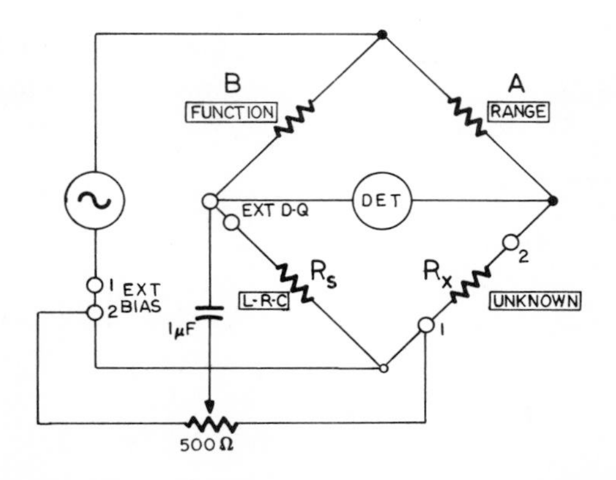

Figure 4.11 Universal Impedance Bridge Circuit for AC Resistance Measurement

OTHER IMPEDANCE MEASURING EQUIPMENT

As we mentioned at the beginning of this chapter, other instruments are available for more precise measurement of individual components of impedance. One of the most important of these uses the Schering bridge.

Schering Bridge

The Schering bridge is somewhat similar to the Maxwell bridge, but is much better for measurement of impedances at radio frequencies. As shown in Figure 4.12, it uses variable capacitors instead of rheostats to achieve a balance. These are air-dielectric precision capacitors, heavily shielded to avoid stray capacitance, radiation and AC pickup.

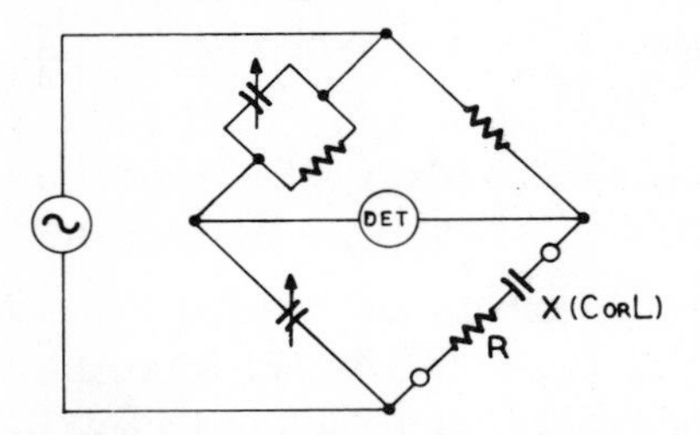

Figure 4.12 Schering Bridge Circuit

The best-known commercial version of this bridge is the *RX Meter*, with a frequency range from 0.5 to 250 megahertz. You set the frequency of the internal oscillator first, and then balance the bridge without anything connected to the input terminals.

You then connect the unknown impedance to the input terminals and rebalance the bridge. The new settings of the resistive and reactive dials show the change caused by the unknown impedance, so the values for the unknown are the difference between the readings with nothing connected and those with the unknown impedance connected.

Q Meter

Q meters are used for direct reading of Q of coils and circuits. From the latter can be calculated distributed capacitance, effective inductance and many other characteristics. Figure 4.13 shows the general principle of a Q meter. An RF oscillator is connected in series with the inductance to be measured and a variable capacitor. The capacitor is adjusted until the circuit is resonant at the frequency to which the oscillator has been tuned.

A built-in VTVM responds to the reactive voltage across the capacitor, indicating it on the meter dial as a Q value.

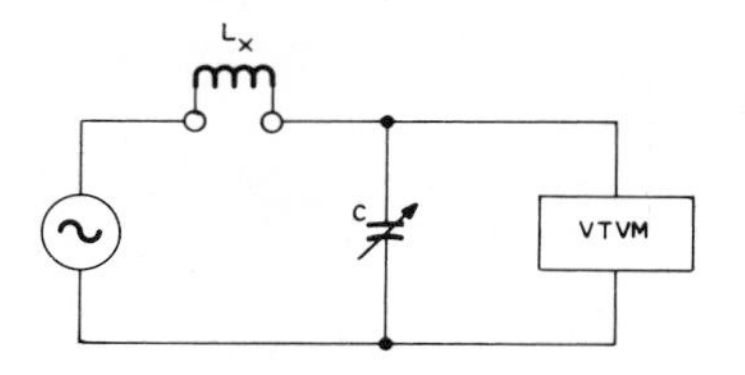

Figure 4.13 Q Meter (Block Diagram)

Phase-Angle Measurement

Measurement of phase angle is done by comparing the signal's phase angle with that of a reference voltage. In Figure 4.14, an oscillator applies a signal to the unit under test (which might be a transformer, an amplifier, or some other circuit) and to the reference input of the phase-angle meter. The output signal from the unit under test is then applied to the signal input of the meter, which indicates the difference in phase between this signal and the reference signal.

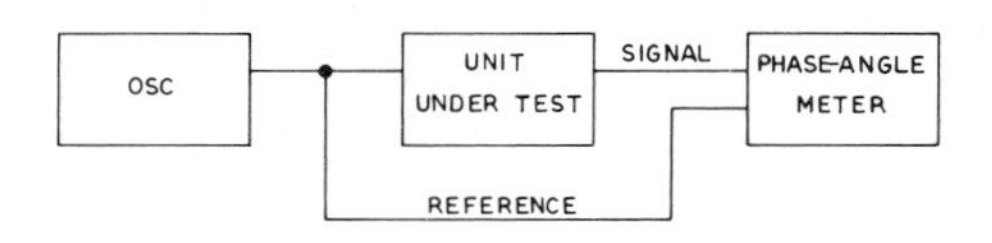

Figure 4.14 How a Phase-Angle Meter Is Used

One way of doing this is to convert both inputs to square waves of equal amplitude, and apply them to opposite sides of a flipflop. If there is no phase difference, the flipflop will not turn on. If there is a phase difference, the flipflop will be on for a length of time proportionate to the difference. Its output causes a corresponding deflection on a meter calibrated in degrees.

Another method is to take half of the reference signal and phase-shift it exactly 90 degrees. There are now two reference signals, the "in-phase" and the "quadrature." You remember that in Chapter 1 we saw that any signal could be considered as consisting of component vectors at right angles. The phase angle is then proportional to their voltages. When the corresponding vectors of the input signal are bucked against the in-phase and quadrature reference signals, the amount by which each vector differs from the reference value will bear a ratio to the other in proportion to the phase angle.

This is the method used in the *complex-ratio bridge*, in which the values of in-phase and quadrature reference voltage required to balance the input signal component vectors are indicated on the in-phase and quadrature voltage decade dials. The phase angle can be calculated from:

$$\tan \phi = \frac{\text{Quadrature reference voltage}}{\text{In-phase reference voltage}}$$

The measurement of impedance brings us to the end of test equipment used for passive components such as resistors, capacitors and coils. The reaction of these to AC is called impedance which, as we have seen, is composed of reactive and resistive components that may be regarded as being in parallel or in series. The most widely-used instrument for measuring impedance is the universal impedance bridge, but other instruments are also available for measuring the individual components of impedance. As a general rule, a specialized instrument will be more accurate than a universal one. A good-quality inductance bridge can have an accuracy of 0.05 percent, a capacitance bridge 0.01 percent, whereas the accuracy of universal impedance bridges is from 0.1 to 1 percent.

5

TESTING VACUUM TUBES, TRANSISTORS AND DIODES

Electronic components are either active or passive. The passive ones are tested by measuring the resistance or impedance they oppose to direct or alternating current, as you saw in the last two chapters. Active components (except for diodes, in some cases) must have DC supply voltages to enable them to function.

Active components are vacuum tubes, transistors and diodes. They are used in circuits mainly to preform the following tasks:

Amplification	AC input signal controls DC supply current so that external power is used to produce an enhanced replica of the input signal.
Oscillation	AC feedback signal controls the DC supply current so that external power is used to sustain oscillation.
Modulation	Two AC input signals control DC supply so that external power is used to produce an output signal consisting of combined replicas of both inputs. (Subsequent circuits remove unwanted signals.)
Demodulation	Modulated input signal controls DC supply so that external power is used to produce an output signal that is a replica of the wanted component of the modulated input. (However, diodes can also do this without using external power.)
Switching	Input DC signal (voltage or pulse) is used to switch DC supply on or off.
Display	Input AC signal controls DC supply to produce visible image.

Photoelectric Incident light controls DC supply to produce electrical signal.

In most cases these functions involve the control of DC by AC. A complete test of such a device would require the application of both DC and AC, but for some tests DC alone is sufficient.

VACUUM TUBE TESTING

There are six things generally tested in vacuum tubes, and four ways of doing so, as shown in Table 5.1.

TABLE 5.1 VACUUM TUBE TESTS

TEST TO BE PERFORMED	METHOD OF TESTING			
	Filament Tester	Emission Tester	Dynamic Tester	Substi-tution
Open heater filament	X	X	X	X
Heater-cathode leakage		X	X	X
Short-circuits between elements		X	X	X
Excessive gas			X	X
Cathode emission		X		X
Mutual conductance			X	X

Filament Tester

A filament tester can only indicate an open heater filament. It consists of a small box with a panel, on which are mounted a set of tube sockets to accommodate the different types of receiver tubes. The heater pin terminals are all connected in parallel across the AC line, but one side is connected through a neon bulb and series resistor. When a tube with a good heater filament is plugged in it completes the circuit, and the bulb lights. If the filament is open the bulb does not light.

You can see that this simple device is similar to the neon tester of Chapter 2. It is helpful to the serviceman checking a series-string of tubes in a television set, since when one goes out they all go out, and it takes time to go through all those tubes with any other tester. Consequently, many servicemen carry filament testers in their tool boxes, although this is their only function.

Emission Tester

A vacuum tube operates by means of a controlled flow of electrons from the material coating its cathode. If the material is worn out or its temperature is too low, it cannot release enough electrons for efficient operation. An emission tester measures this capability, and tells you if the electron flow is normal, diminished or absent.

The simplified circuit for emission testing is shown in Figure 5.1. When P1 is plugged into a service outlet and switch S1 is closed, 40 to 50 AC volts appear across one secondary of T1 and 117 AC volts across the other. A number of taps on this secondary provide all the voltages, from 117 down to 1.5, used for tube heater filaments. Before plugging your tube into the socket XVI you select the correct heater voltage by turning the selector switch S2 to the proper indication.

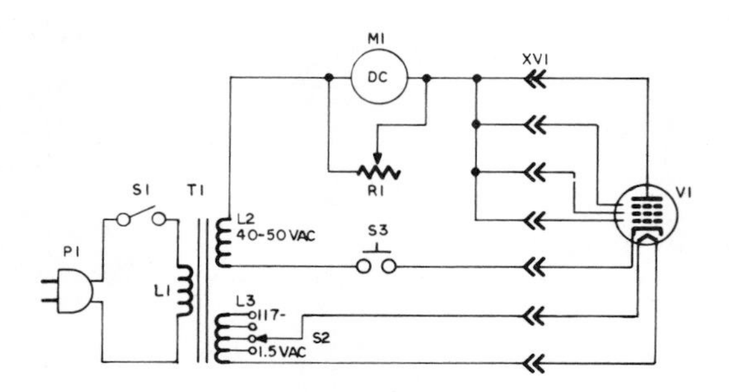

Figure 5.1 How an Emission Tube Tester Works

When the tube has had time to warm up to its normal operating temperature (at least 30 seconds), you rotate R1 to the position given for that tube in the tester's instruction booklet, and depress the push-button switch S3. This places 40 to 50 volts AC across the tube. All the elements of the tube except the cathode and filament are connected together, so that they all become positive every other half-cycle of the AC voltage, attracting electrons from the negative cathode. In other words, the tube is operated as a diode, and a train of DC pulses at the 60-hertz rate flows through the meter M1. The sensitivity of this meter was adjusted for the tube when you set R1. Its dial gives a simple GOOD—?—REPLACE indication according to whether the electron flow is normal, borderline or weak. No indication at all might mean an open heater filament or cathode lead.

A tube that tests GOOD on this tester may still not work in a circuit, because the test was not made with normal operating voltages applied, but one that reads in the REPLACE section, even if working in its circuit, has only a very short life left at most.

Dynamic Tester

The dynamic tester, as shown in Figure 5.2, sets up a standard circuit for the tube being tested, with normal operating voltages and load, so that the meter can indicate actual performance. A small AC signal is applied to the tube's grid, and the meter reads the AC output.

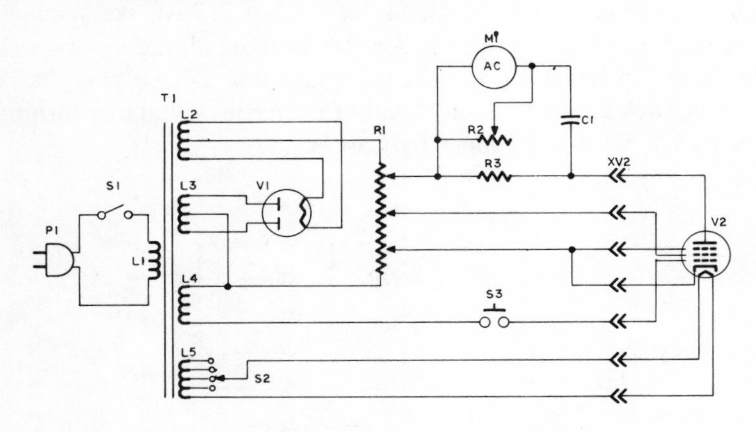

Figure 5.2 How a Dynamic Tube Tester Works

The heater filament voltage is selected as before, using S2. However, a full-wave rectifier is now included to give a DC supply across R1, from which operating voltages are obtained for the electrodes of the tube under test.

The meter M1 is an AC type (PMMC with rectifier) that reads the AC output signal across the load resistor R3. Its sensitivity is adjusted by R2, and DC is blocked by C1. This output signal appears when S3 is closed to apply a small AC signal to the grid of V2.

The meter on a dynamic tester also shows a GOOD—?—REPLACE indication, but reads transconductance in micromhos as well. For most tests, the GOOD—?—REPLACE indication is enough, but the micromho scale enables you to select matched pairs of tubes for push-pull circuits, or a tube with a specified transconductance for a critical circuit.

Emission and Dynamic Testers Compared

In a dynamic tester, the tube is operated with normal electrode voltages, so a real test of its amplifying capability is given. However, this is done at only 60 hertz, so it tells you nothing about performance at VHF or UHF, where conditions such as excessive interelectrode capacitance, which would have no effect at low frequencies, might render it useless.

An emission tester saves time because it is simpler. Also, it can give a better test for power output tubes which require a high cathode current. The dynamic tester does not give this type of tube a full workout. The emission tester is also much less expensive.

Neither tester can give a good test to oscillators, so that for these and for tubes used at high frequencies, the best test is by substitution.

Substitution

Substituting a known good tube for the suspected faulty one is the only test that can pinpoint the tube as the cause of failure under *all* circumstances. However, this method runs into difficulties when more than one tube is defective in an instrument being tested. Also, it doesn't tell you what the problem was. And finally, you need to keep a large stock of tubes on hand, which is expensive. Consequently, a tube tester should normally be used, but you should also bear in mind that some tubes in some circuits can be tested *only* by substitution. A good example is in the "front end" or tuner section of a TV set. All channels up to 5 come in good, but 7 and up are missing. The RF amplifier tube tests GOOD on the tube tester. After sustituting a new tube, all channels are received. The old tube had developed excessive interelectrode capacitance, which was grounding the higher frequencies.

Other Tests

Both emission and dynamic testers also test for filament continuity, shorts between electrodes and heater-cathode leakage. Dynamic testers check for gas in the tube envelope as well. Both testers test diodes.

FILAMENT CONTINUITY

The test for filament continuity is essentially the same as in the filament tester already described, except that depressing a pushbutton places the meter in series with the filament, so that the pointer drops to zero if the filament is open. Also, the proper filament voltage and pin connections are selected for the test.

SHORT-CIRCUIT AND LEAKAGE TESTS

Figure 5.3 shows a type of circuit used for testing for shorts between electrodes. When a good tube is plugged in, all five neon bulbs light, but if a short exists between two or more of the tube's electrodes, the corresponding bulbs will be bypassed and will not light. If a short exists between the cathode and control grid, for instance, bulb 4 will not light. If the control grid and screen grid are shorted together, bulbs 2, 3 and 4 will not light.

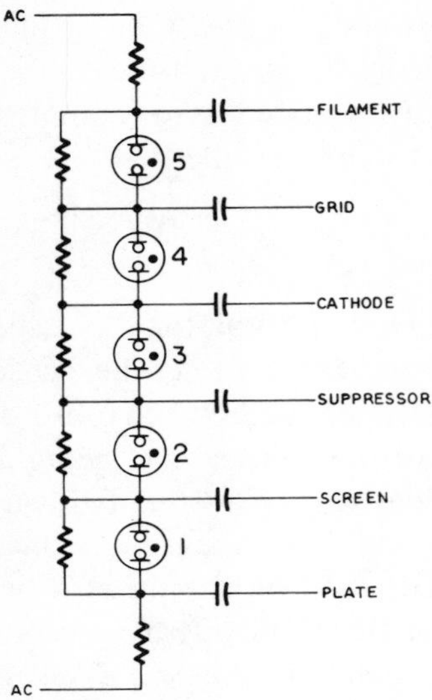

Figure 5.3 Circuit for Testing for Shorts between Electrodes

This test circuit requires no controls to be set, and takes place automatically every time a tube is plugged in to a socket, so you don't have to remember to do it before doing other tests. Some testers, however, use switches or pushbuttons to test each element in turn. In this case, only one neon bulb is required, and the position of the switch tells you which element is shorted.

This test usually checks for heater-cathode leakage as well, but sometimes a separate test is provided by connecting the DC plate voltage between plate and filament, with the cathode open. If the meter indicates DC flowing under these conditions, there must be leakage from the heater filament to the cathode.

GAS TEST

If gas is present within the tube envelope, its molecules collide with electrons on their way from the cathode to the plate, and become ionized. The positive ions collect on the control grid because of its negative bias. As a result, the grid-cathode bias is reduced, resulting in excessive plate current and distorted output.

The best circuit for this test is one that applies normal operating voltages to the tube, but in addition connects a resistor (100 to 200 kilohms) to the grid to allow the positive charge to escape to ground. This current also passes through the meter. If it causes the pointer to deflect more than a very small amount, the tube is gassy.

It is also possible to have a DC meter connected in the plate circuit. If the reading increases considerably when the grid resistor is connected, gas is indicated.

Using a Dynamic Tube Tester

Figure 5.4 illustrates a well-known dynamic tube tester. The procedure for using it is as follows:

1. Connect the power cord to a suitable service outlet.
2. Turn the POWER LINE ADJUST control to ON, and adjust it to place the meter pointer exactly over the LINE TEST mark. This compensates for variations in the line voltage.
3. Rotate the roll chart to get the data for the tube you wish to test.

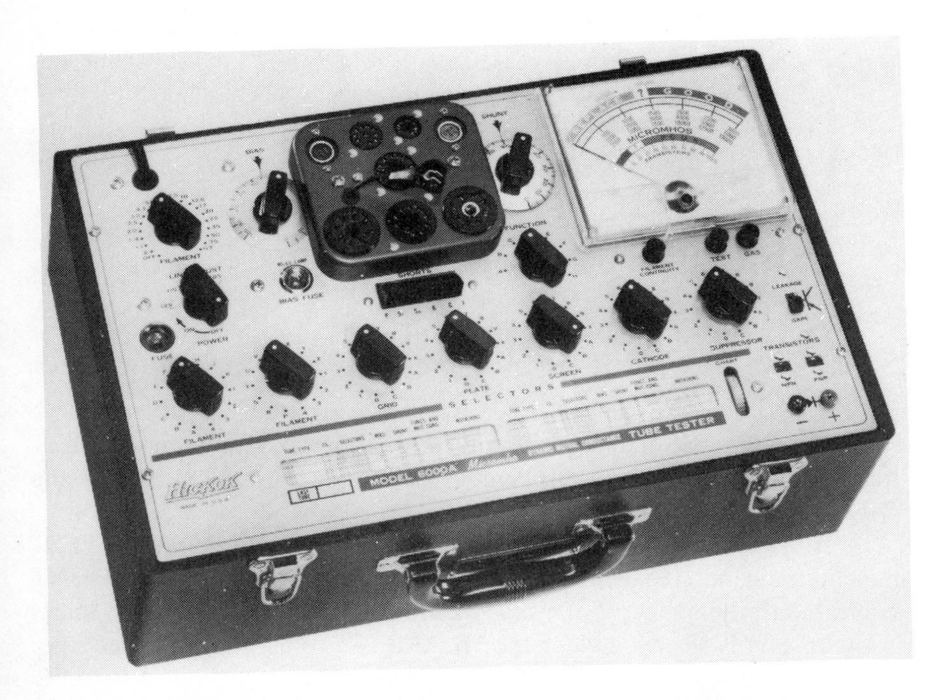

Figure 5.4 Dynamic Tube Tester that can also check diodes and transistors. *(Courtesy Hickok Electrical Instrument Co.)*

4. Set the controls as shown in the roll chart. For instance, set the FILAMENT control in the top left-hand corner of the panel in accordance with the voltage given in the FIL column. Then set the SELECTORS as instructed in the next column. For example, if this says JR-6237-5, set the SELECTOR knobs as follows:

FILAMENT	FILAMENT	GRID	PLATE	SCREEN	CATHODE	SUPPRESSOR
J	R	6	2	3	7	5

   This will connect the operating voltages to the correct base pins of the tube under test.

5. Set the BIAS and SHUNT controls as shown in their respective columns. This will apply the proper bias to the tube and adjust the meter sensitivity to give the proper reading of the output.
6. Set the FUNCTION control in accordance with the FUNCTION column. This selects the proper test circuit for the tube under test.
7. Now insert the tube in its proper socket and allow it at least 30 seconds to warm up.
8. Check continuity of the heater filament by depressing the FILAMENT CONTINUITY pushbutton. If the filament is open, the meter pointer will go to zero, otherwise stay at or near LINE TEST.
9. Look at the five glow lamps under the lamp shield (center of panel). All five should glow. If any are dark, there is a short in the tube, and it must be discarded. To make sure there are no intermittent shorts, tap the tube lightly with your fingernail while watching the bulbs.
10. Check that all SELECTORS have been set correctly, and depress the TEST pushbutton. The meter will indicate the quality of the tube on the REPLACE—?—GOOD scale.
11. To read mutual conductance, check the SHUNT control setting, and read from the corresponding MICROMHOS scale (there are three ranges).
12. To check for gas, depress the TEST and GAS pushbuttons together. If the meter pointer indicates more than 2 microamperes (two minor divisions), the tube is gassy.

Rectifiers and diodes are tested for emission only. When the TEST pushbutton is depressed the meter pointer must deflect above the DIODES OK mark.

To ascertain the life expectancy of the tube, hold down the TEST pushbutton as in step 10 above, and at the same time adjust the SHUNT control until the meter pointer reads 2000 on the 0-3000 micromho scale. As soon as it is steady, reduce the FILAMENT voltage selector one step. If the pointer remains in the GOOD section the tube still has a long life expectancy.

TRANSISTOR TESTING

In testing transistors, we are usually interested in only two basic characteristics, measured under DC conditions. These are *gain* and *leakage*.

In a vacuum tube gain is the ratio of output voltage to input voltage (V_{out}/V_{in}). In a transistor, it is the ratio of output current to input current (I_C/I_B in the common-emitter circuit). This is called the *forward current transfer ratio*, or *beta* (β). It can be AC (h_{fe}) or DC (H_{FE}). Transistors generally use DC. Figure 5.5 shows a typical circuit for measuring gain.

Collector-base leakage (I_{CBO}) is measured by applying a voltage across the junction, with the emitter open. This is really a diode leakage test, as we shall see under diode testing. The circuit is shown in Figure 5.6.

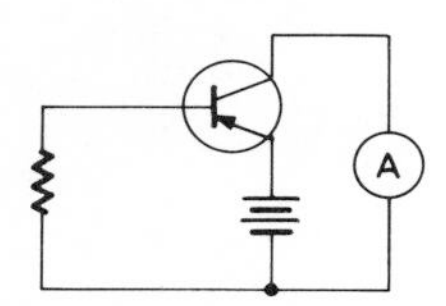

Figure 5.5 Circuit for Measuring H_{FE}

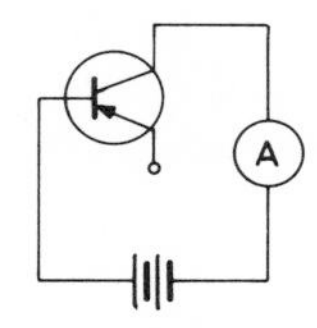

Figure 5.6 Circuit for Measuring I_{CBO}

These circuits may be combined, as shown in Figure 5.7, where a switch enables you to change from one circuit to the other.

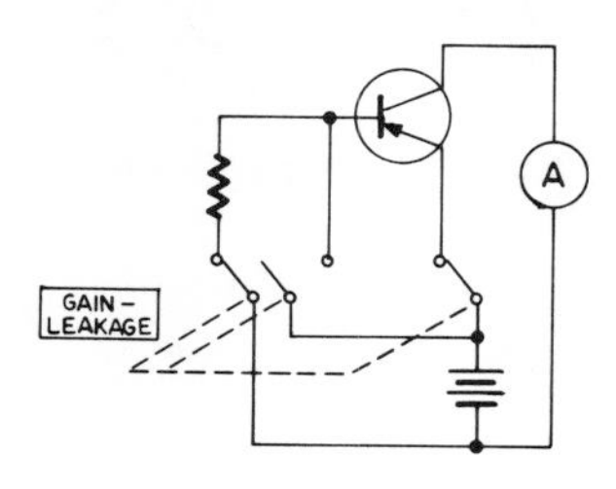

Figure 5.7 How a Transistor Tester Works

In the tester shown in Figure 5.4, provision is made for performing these transistor tests as well as tube tests. Sockets for NPN and PNP transistors are located in the lower right-hand corner of the panel. To test a transistor you proceed as follows:

1. After connecting the tester to a power outlet, rotate the POWER LINE ADJUST control until the meter pointer is over the LINE TEST mark.

2. Set the FUNCTION switch to H. The meter pointer will return to zero.
3. Insert the transistor into the proper socket (check to be sure whether it is NPN or PNP).
4. Adjust the SHUNT control until the meter reads 100 percent on the lower scale. If the pointer fails to deflect upscale, the transistor is defective.
5. Push the LEAKAGE-GAIN slide-switch to the LEAKAGE position. If the meter pointer is in the GOOD area on the lower scale, the transistor leakage is within allowable limits, but if it is in the POOR area, it should be discarded.

Other tests are possible with more elaborate equipment. The *semiconductor curve tracer* is an oscilloscope (see Chapter 9) designed for displaying the characteristic curves of transistors. The oscilloscope applies a sweep voltage to the collector and a step voltage to the base or emitter, whichever is ungrounded. This results in a display like that shown in Figure 5.8, from which the engineer can verify if the device meets its specifications, or predict its performance in a circuit. The curve tracer also tests diodes and integrated circuits in the same way.

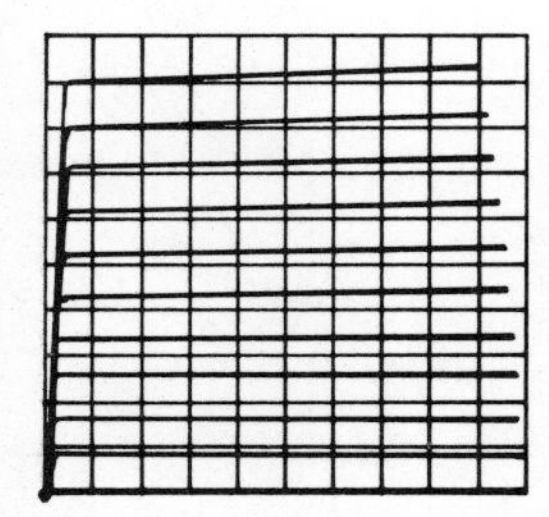

Figure 5.8 How a transistor family of curves looks on the screen of a curve-tracer oscilloscope

TESTING SEMICONDUCTOR DIODES

Semiconductor diodes include silicon, germanium, selenium and copper oxide rectifiers. In checking vacuum-tube diodes, we were concerned with emission, but with these the important consideration is leakage. Semiconductor diodes are one-way resistors: high resistance in one direction, low in the other. Some current flow in the high-resistance direction is to be expected, but as long as it is low enough to be negligible the diode is good. Consequently, an I_{CBO}-type test, as shown in Figure 5.6, is a good diode test.

The tester in Figure 5.4 has a red (+) and black (−) jack near the transistor test sockets in the lower right-hand corner of the panel. To check a diode:

1. Connect the diode's positive (cathode) lead to the black (−) jack, negative lead to the red (+) jack. This places forward bias on the diode.
2. Set POWER LINE ADJUST control so that the meter pointer is over the LINE TEST mark.
3. Set the SHUNT control to 100.
4. Set the FUNCTION switch to H. Meter pointer will return to zero.
5. Adjust SHUNT control until meter pointer indicates 100 percent on lowest scale.
6. Reverse the diode leads, so that the positive (cathode) lead is now connected to the red (+) jack, the negative to the black (−) jack. This places reverse bias on the diode.
7. The meter pointer should indicate less than 10 percent if the diode is good. A video detector diode, however, can read 10 percent and still be good.

Diodes can, of course, be tested in the same way with an ohmmeter. Just connect the diode first one way, then the other, without changing ranges. The lower reading must not exceed one-tenth of the higher reading. However, this is a rather more rough-and-ready method than using the tester, because of the fixed battery voltage of the ohmmeter.

Checking active components is a steppingstone between the testing of passive components and the evaluation of circuit performance. It does not require you to be familiar with all the thousands of vacuum tubes, transistors and diodes, but you should know enough about their main categories to understand the meaning of the tests you perform on them. You'll find that the information given in tube and semiconductor manuals will be very helpful to you in this respect.

6

SPECIAL USES FOR MULTIMETERS AND DIFFERENTIAL VOLTMETERS

We now leave the testing of component parts, such as resistors, capacitors, coils, vacuum tubes and transistors, and move into the area of functional testing. In this area we are concerned with the performance of groups of components that together form amplifiers, oscillators or other circuit assemblies.

Sometimes you'll want to check the performance of a circuit to make sure it is doing what it should do, making adjustments to improve this if necessary. Or else you may want to track down the cause of a breakdown in its operation, so that you can correct it. The first requirement comes under the heading of maintenance and calibration, the second is called troubleshooting.

The characteristics that reveal circuit behavior are, first, the output of the equipment of which the circuit is a part. The radio does not play, the television picture rolls, the hi-fi hums. These symptoms often point to the circuit involved. The circuit itself may be built around a vacuum tube, which is then the prime suspect. But if this is not the cause of the trouble you will then check other circuit characteristics, which are voltages at various points, and waveforms. Current is not measured as often, because to do so usually means opening the circuit, as you saw in Chapter 2; it can be calculated, in most cases, by measuring the voltage drop across a resistor. In this chapter, we shall discuss test instruments used mainly for voltage measurement, but since some of them can also measure current and resistance they are called *multimeters*. The principal exception is the *differential voltmeter,* which measures voltage only. Together, these make up the most frequently-used test instruments, classified under the general name of *service meters*.

The four main groups are shown in Table 6.1. They are the volt-ohm-milliammeter, electronic voltmeter (vacuum-tube voltmeter or transistorized voltmeter), digital multimeter and differential voltmeter, arranged in increasing order of accuracy and decreasing order of circuit loading. Not every model makes all the measurements indicated, but typically they do, or may do so by the addition of optional circuitry.

The first two categories are general-purpose, service-shop instruments. Their accuracy is sufficient for that use, but does not compare with the accuracy of the second pair. However, they are more rugged, portable and less expensive. The second pair of instruments are much more elaborate, and therefore, more likely to be found in the laboratory or factory functional test area.

VOLT-OHM-MILLIAMMETER

In Chapter 2, we mentioned that direct-reading voltmeters, voltmeters, ohmmeters and current meters are often combined in one. Figure 6.1 illustrates a typical example of a volt-ohm-milliammeter.

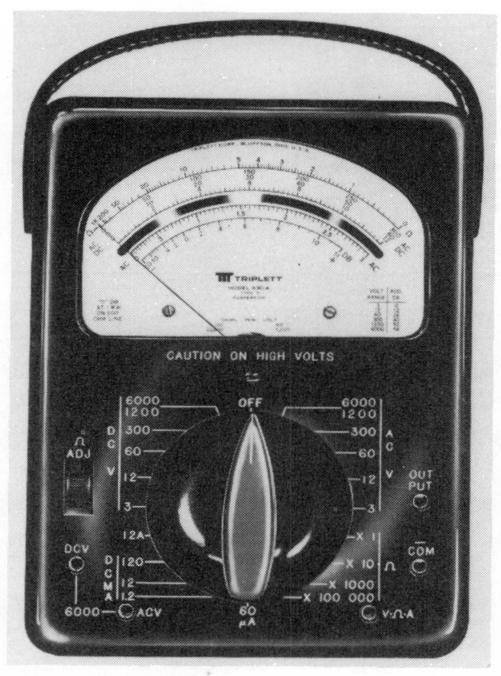

Figure 6.1 Typical Volt-Ohm-Milliameter *(Courtesy Triplett Corporation)*

TABLE 6.1 MULTIMETERS AND DIFFERENTIAL VOLTMETERS.

INSTRUMENT		MEASUREMENT FUNCTION			
		VOLTAGE		CURRENT	RESISTANCE
		DC	AC*		
Volt-Ohm-Milliammeter (VOM)	Accuracy (%)	1.5 - 3	3 - 4	4	3
	Input Impedance	20,000Ω/V	5000Ω/V		
Electronic Voltmeter (VTVM, TVM)	Accuracy (%)	1 - 3	3	-	1 - 5
	Input Impedance	10MΩ	-100MΩ	1 - 10MΩ	
Digital Multimeter (DMM)	Accuracy (%)	.0025 - .1	.05 - .2	-	.01 - .05
	Input Impedance	10MΩ - 10GΩ	1MΩ		
Differential Voltmeter (DVM)	Accuracy (%)	.0025 - .1	.05 - .1	-	-
	Input Impedance	** ∞	1MΩ		

*Accuracy stated is true only for frequency range specified for instrument. This runs from as low as 60 Hz for some VOMs, through 100 kHz for DMMs and DVMs, to several MHz for some VTVMs.

**At null. Depends on model. Some solid-state models are 10MΩ on higher ranges.

The heart of a VOM is a sensitive PMMC movement. In the circuit in Figure 6.2, this is a 50-microampere movement, connected as a DC voltmeter, AC voltmeter, current meter or ohmmeter by setting the function switch. This switch also selects ranges for each function. VOMs have separate scales for DC volts and milliamperes, AC volts and ohms.

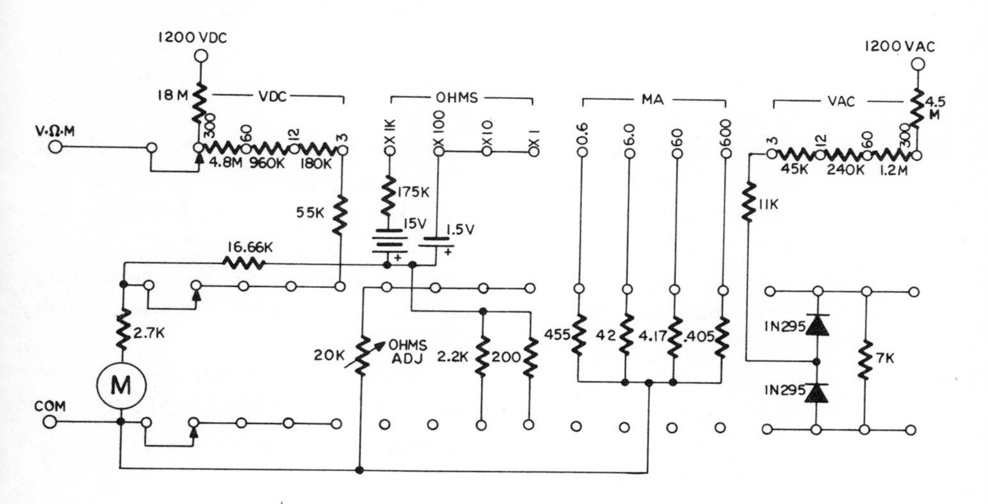

Figure 6.2 VOM Circuit

The VOM can use its meter to measure direct current in a circuit, something that electronic voltmeters cannot do. You saw how to do this in Chapter 2. But the VOM is used far more for making voltage measurements than for anything else.

In making voltage measurements, the loading effect of the meter may have a considerable effect on the reliability of the measurement. All good VOMs have a minimum of 20,000 ohms per volt on DC. But some circuits are already loaded so heavily that the reading of even a good VOM may be meaningless. A more sensitive meter movement is possible but not practical because of its fragility. In this situation, you should use an electronic voltmeter.

The chief advantage of the VOM is its portability. Although battery-operated electronic voltmeters are also portable, only the VOM can be made small enough to slip in the pocket and weigh less than a pound. Some modern VOMs, by using field-effect transistors in the DC circuit (FET-VOMs), are able to give the input impedance of an electronic voltmeter. The FETs are used in a circuit similar to the vacuum-tube bridge described below, and require battery current. However, as this may be as low as 10 microamperes only a small battery is required.

Using Your VOM

As we have already discussed voltage, current and resistance measurements in Chapters 2 and 3, we can skip the actual meter-reading techniques, and summarize the use of a VOM in seven simple steps:

1. Select function (DC voltage, AC voltage, DC current, ohms).
2. Select range. If there is any doubt as to which range to use, start with the highest one, and switch downwards to the one that gives the largest deflection. This will avoid pegging the meter.
3. Connect test leads (in the instrument in Figure 6.1 connect black test lead to COM jack, red to V-Ω-A jack). However, when using the 5000V range, connect red to 5000V jack, and set switch to 300V—DC or AC as applicable.
4. If, before taking a voltage measurement, the pointer does not rest on zero, adjust it by turning the screw below the meter face with a small screwdriver.
5. For resistance measurements, short the leads together first, and set pointer to zero on ohms scale by means of the OHMS ADJ control.
6. If the pointer cannot be zeroed on ohms ranges, either or both batteries must be replaced.
7. Connect leads to points between which measurement is to be made. If meter pointer goes downscale, reverse the leads. When measuring current, do not test directly across any potential as you may burn out the meter or shunt resistor.

CURRENT MEASUREMENT

The VOM uses its PMMC movement with various shunts to measure direct current. The range of current measured depends upon the sensitivity of the meter movement and the values of the shunting resistors, and usually does not exceed 10 amperes. It obviously cannot measure alternating current directly, but must first convert it to DC.

Conversion to DC can be done with an AC clamp ammeter adapter, as illustrated in Figure 6.3A. This device consists of a transformer with resistors shunted across its secondary, as shown in Figure 6.3B. The jaws are opened by pressing the lever at the side, and they are placed around the conductor carrying the current to be measured. When allowed to close, the jaws form the iron core of transformer T1, and the conductor becomes a one-turn primary winding. The secondary winding has many more turns, so that the weak magnetic field produced by a current of six amperes around the single conductor induces a potential drop of 3 volts across R1. This is connected to the input of the VOM, and gives a full-scale deflection on its 3VAC range. The same meter range is used with all the adapter ranges, the current value being calculated as follows:

$$\text{AC Current} = \text{AC Voltage} \times \frac{\text{Adapter Range}}{3}$$

Selection of ranges is done with the adapter switch S1.

Clamp ammeters are also available as separate instruments, and clamp probes as accessories for some VTVMs.

Figure 6.3a AC Clamp Ammeter Adapter mounted on a pocket VOM *(Courtesy Triplett Corporation)*

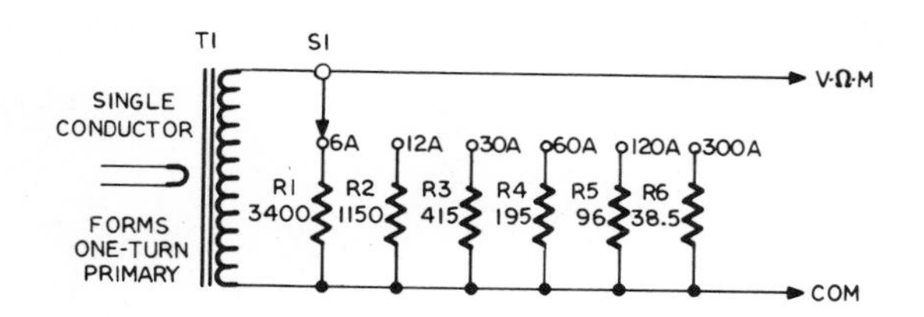

Figure 6.3b How the Clamp Ammeter Adapter Works

ELECTRONIC VOLTMETERS

Electronic voltmeters are vacuum-tube or transistorized voltmeters that require a power source for operation and give an analog output by means of a meter reading. Digital voltmeters operate by counting and differential voltmeters by potentiometric measurement, so although they are also electronic instruments, we shall deal with them separately.

Vacuum-Tube Voltmeters

The sensitivity of direct-reading multimeters as voltmeters cannot for practical reasons exceed 20,000 ohms per volt. The vacuum-tube voltmeter is more sensitive because the current through the meter coil does not come from the circuit under test but from the VTVM's own power supply.

Figure 6.4 shows the principle of a DC VTVM circuit. You can see it is really a Wheatstone bridge in which the resistances in the arms have been replaced by vacuum tubes. V1 has taken the place of the "unknown" resistance and V2 has become the standard. The grid of V2 is grounded, so that the plate resistance of this tube remains constant. V1's grid is connected to the voltage being measured, so its plate resistance decreases as this voltage becomes less negative. At zero volts input the two tubes conduct equally, so their plate voltages are the same, and no current flows through the meter. When making a measurement, V1's grid goes more positive in proportion to the positive voltage measured, so it conducts more and its plate voltage falls. Now that this is no longer equal to the voltage on V2's plate, current flows through the meter movement. The deflection of the meter pointer is proportionate to the voltage applied to V1's grid.

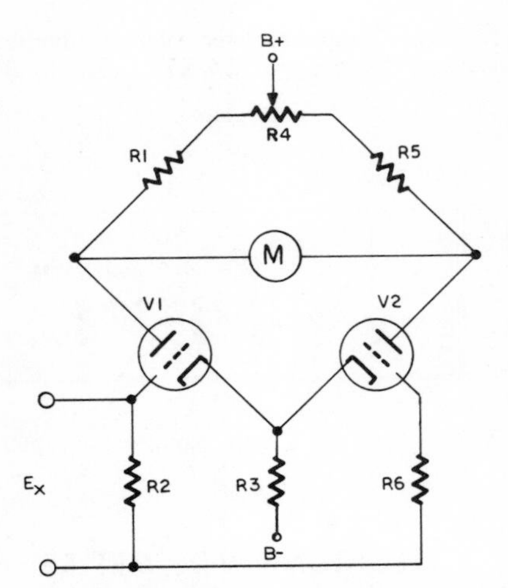

Figure 6.4 A Vacuum-Tube Voltmeter is a Wheatstone Bridge

Resistor R4 is a balancing adjustment to compensate for minor differences between V1 and V2. In some circuits it may be in the cathode circuit instead of the plate circuit. By adjusting it until the meter reads zero, you make the plate voltages of V1 and V2 equal when no voltage is applied to V1's grid.

V1 and V2 are operated as Class A amplifiers, so that their output is linear and input impedance is extremely high. They therefore have a negligible loading effect.

Figure 6.5 illustrates a typical VTVM, and Figure 6.6 gives its schematic. You can see it has the identical bridge circuit of Figure 6.4, with the addition of the input circuits required to enable it to measure DC, AC and ohms. The power supply and calibration controls are also shown. We shall deal with each of these functions separately.

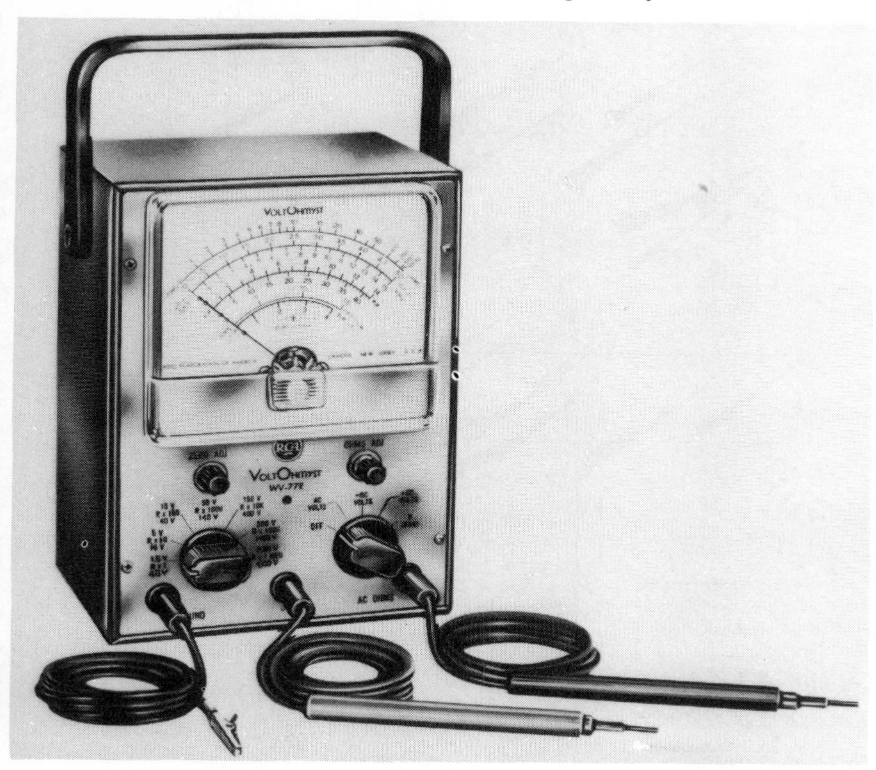

Figure 6.5 Typical VTVM *(Courtesy RCA Corporation)*

DC VOLTAGE MEASUREMENT

Figure 6.7 shows the bridge and input circuits for DC measurement. The main feature is the *attenuator* consisting of resistors R27 through R33. These are mounted on the range switch S1, section 1 (rear), in such a way that rotation of the switch connects the grid of V2A via R9 in turn

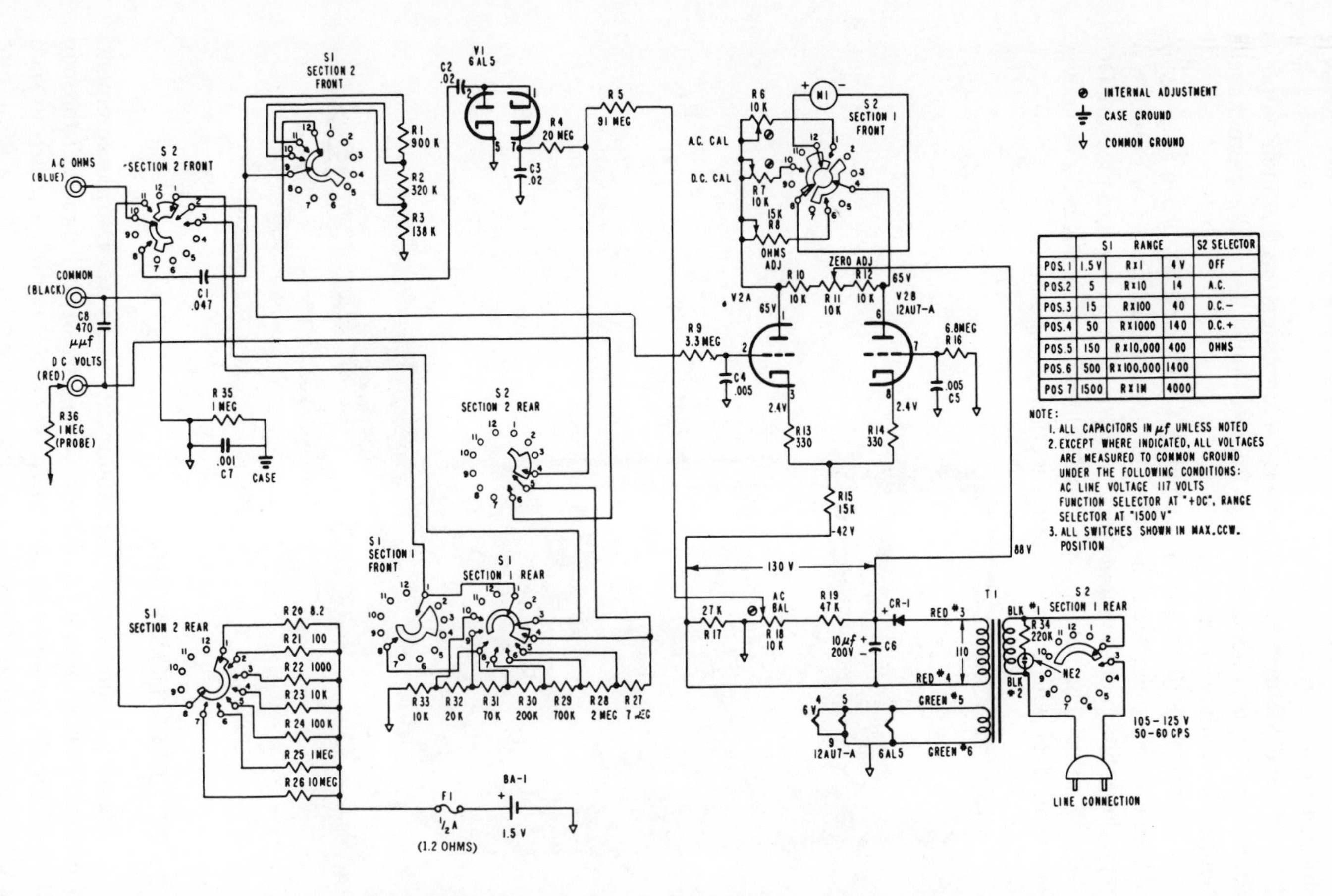

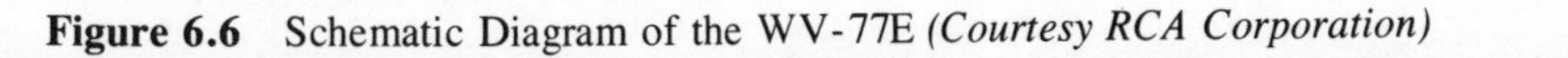
Figure 6.6 Schematic Diagram of the WV-77E *(Courtesy RCA Corporation)*

to each of the positions numbered 4 through 10, according to the range selected. This voltage divider has a total resistance of 10 megohms. The probe used for DC measurement also contains a 1-megohm resistor R36 in series with the divider, so that the total input impedance for DC is 11 megohms. This has a very small loading effect (negligible for most purposes) on the circuit where the measurement is being made.

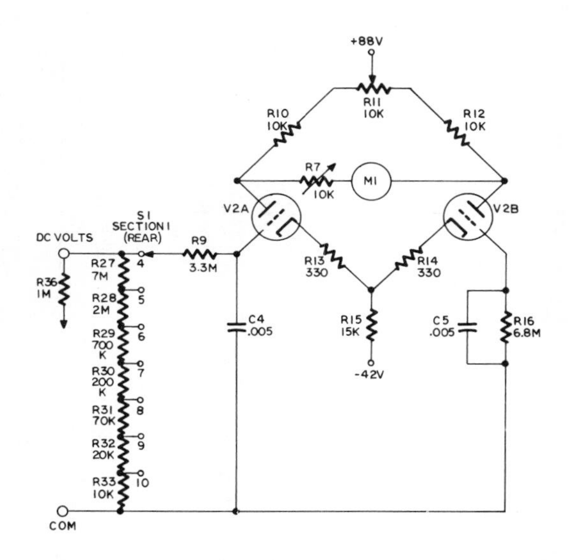

Figure 6.7 VTVM Circuit for DC Voltage Measurement. (This is a re-arrangement of part of the circuit in Figure 6.6.)

The meter is calibrated by adjusting R7 so that an input of 1.5 volts gives a full-scale deflection of the pointer with S1 at position 4. The actual potential at this position will be 1.36 volts because of the fraction dropped across R36.

To measure higher voltages, the range switch is rotated to the position that divides the potential at position 4 by the factor required to ensure that the potential applied to V1A's grid does not exceed 1.36 volts. This is achieved by choosing voltage-divider resistance values so that each position of S1 selects the proper attenuation factor to make the voltage to be measured look to the meter like an input of between 0 and 1.5 volts, regardless of its real value, as in Table 6.2.

The resistor R36 in the probe also serves the purpose of decoupling the VTVM from the circuit under test, so that the probe does not detune the circuit. The low-pass filter consisting of R9 and C4 reduces the effect on V2A's grid of stray pickup from the power line, or RF from signal generator or other equipment nearby.

TABLE 6.2 VTVM ATTENUATOR RESISTANCE VALUES FOR RANGE SWITCH

S1 METER POSITION	RESISTANCE TO GROUND (MΩ)	RANGE (VOLTS)	X	ATTENUATION FACTOR (MΩ)	=	METER SEES (VDC)
4	10	1.5	X	10/10	=	1.5
5	3	5	X	3/10	=	1.5
6	1	15	X	1/10	=	1.5
7	.3	50	X	.3/10	=	1.5
8	.1	150	X	.1/10	=	1.5
9	.03	500	X	.03/10	=	1.5
10	.01	1500	X	.01/10	=	1.5

The function switch has positions for −DC VOLTS or +DC VOLTS. When measuring a negative voltage, the grid of V2A will become more negative in proportion to the voltage, causing the meter current to flow in the opposite direction, and giving a downscale reading. To prevent this, S2, section 1 (front), reverses the connections for negative voltage measurements. This part of S2 is shown in Figure 6.6, in the meter circuit.

AC VOLTAGE MEASUREMENT

The circuit for measuring AC voltage is shown in Figure 6.8. When the function switch is set to AC VOLTS, the DC input is disconnected and the AC OHMS input is connected by S2, section 2 (front), to the voltage divider R1 through R3, via C1, a DC-blocking capacitor. S1, section 2 (front), connects the AC voltage via C2 to V1 which, with C3 and R4, forms a full-wave rectifier circuit. S2, section 2 (rear), then connects the DC output of this circuit to the attenuator R27 through R33, and via R9 to V2A.

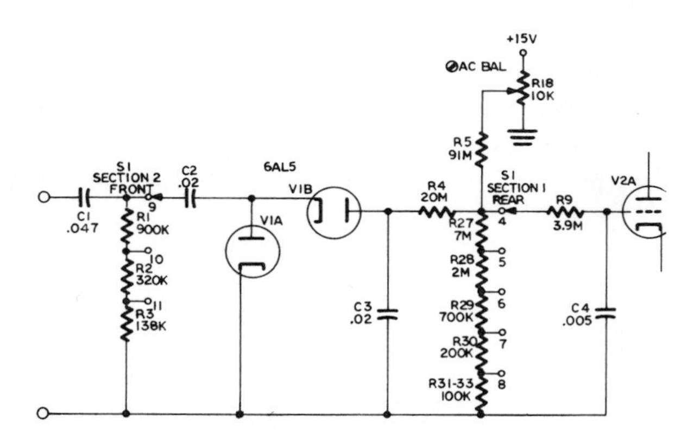

Figure 6.8 VTVM Circuit for AC Voltage Measurement. (This is a re-arrangement of part of the circuit in Figure 6.6.) When S1, Section 1, Rear, is at position 8, S1, Section 2, Front, is at its position 9. With further rotation of S1, R9 remains connected to position 8, while S1, Section 2, Front, moves to position 10 and then to position 11.

The loading effect of the AC input on the circuit under test is greater than that of the 11-megohm DC input because R1, R2 and R3 total only 1.358 megohms. However, this is not a serious matter for most AC measurements. A separate voltage divider ahead of the attenuator is necessary if the meter is to read the correct value of an AC voltage.

The meter dial has scales for both RMS and peak-to-peak values, but readings will only be within the specified accuracy if the measured voltage has a sinusoidal waveform (see Chapter 2).

A vacuum-tube rectifier is used in preference to a copper-oxide or other semiconductor diode because it has better frequency and voltage ranges. However, a small current always flows between the cathode and plate of a vacuum-tube diode whenever the cathode is heated. This current would affect the meter reading, except that the two halves of V1 are connected with opposite polarity, so that the currents through each would cancel out if they were exactly equal. As this is seldom the case, the AC BAL control R18 is provided to eliminate the difference. An internal AC CAL adjustment R6 is also provided to calibrate the meter for AC.

RESISTANCE MEASUREMENT

Unlike the ohmmeter circuits of Chapter 3, a VTVM does not measure resistance directly. The range selector places one of the standard resistors R20 through R26 in series with the resistance being measured, as shown in Figure 6.9. The current from the battery flowing through both gives a potential drop across the standard resistor according to the ratio it bears to R_x. If they happen to be equal, half the battery voltage appears at the grid of V2A, resulting in a half-scale deflection. When the input is shorted, V2A's grid is grounded (through R9), so that it is at the same potential as V2B's grid, and there is no meter deflection. When it is open, the full battery potential appears on V2A's grid, and the meter reads full scale.

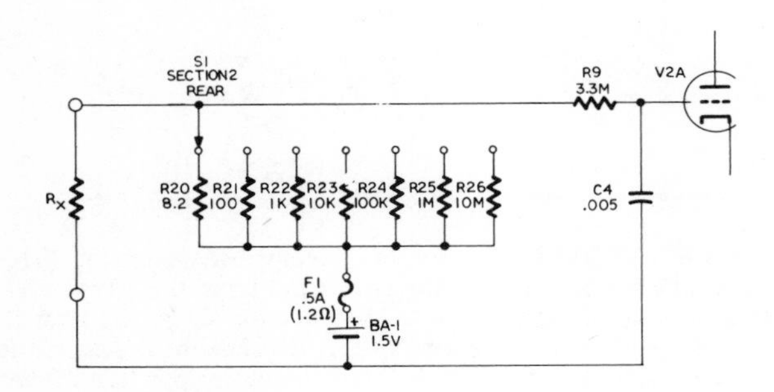

Figure 6.9 VTVM Circuit for Resistance Measurement. (This is a re-arrangement of part of the circuit in Figure 6.6.)

An OHMS ADJ control in the meter circuit is provided to compensate for aging of the battery. When using the VTVM for resistance, first short the AC-ohms and common leads together and adjust the ZERO ADJ control to zero the meter pointer if necessary; then disconnect the leads and adjust the OHMS ADJ control to bring the pointer over the extreme right-hand mark on the ohms scale. If excessive readjustment is required on switching from high to low ranges or vice versa, the battery may need to be replaced. To check its condition, set the range selector to R X 1 and the function switch to OHMS. Adjust the OHMS ADJ for a full-scale deflection, then short the probe and ground clip together for about 10 seconds. If the meter pointer does not return reasonably promptly to full scale when the leads are disconnected, the battery is weak and should be replaced.

You should never attempt to measure resistance in any circuit while it has power on. However, if you should do this accidentally with this VTVM the range resistors would be protected by the fuse F1.

POWER SUPPLY

The power supply for this VTVM (see Figure 6.6) is a simple half-wave transformer type, with outputs of +88 VDC, −42 VDC and 6.3 VAC. The power switch is part of the function switch S2. A neon lamp lights when it is on.

METER SCALES

The meter dial has seven scales, as shown in Figure 6.10. These are used as in Table 6.3. The zero-center scale allows you to use the VTVM as a null meter, after setting the pointer to the center zero by adjusting the ZERO ADJ control.

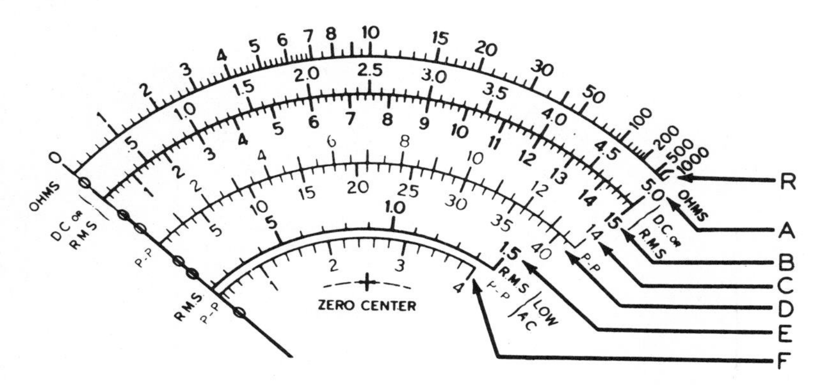

Figure 6.10 VTVM Meter Scales *(Courtesy RCA Corporation)*

USING YOUR VTVM

Using a VTVM is similar to using a VOM, except that with a VTVM you have to plug in the power cord to a service outlet and turn the power on. As the tubes warm up they will often show a little temporary

TABLE 6.3 VTVM METER SCALES
(See Figure 6.10)

RANGE			DC SCALES	AC SCALES (RMS)	(P-P)	OHMS SCALE
1.5 V	4 V	R X 1	B X .1	E X 1	F X 1	R X 1
5 V	14 V	R X 10	A X −1	A X 1	C X 1	R X 10
15 V	40 V	R X 100	B X −1	B X 1	D X 1	R X 100
50 V	140 V	R X 1000	A X 10	A X 10	C X 10	R X 1000
150 V	400 V	R X 10K	B X 10	B X 10	D X 10	R X 10,000
500 V	1400 V	R X 100K	A X 100	A X 100	C X 10	R X 100,000
1500 V	4200 V	R X 1 MEG	B X 100	B X 100	D X 100	R X 1,000,000

unbalance by a deflection of the meter pointer, regardless of range. The pointer gradually returns to zero as the tubes reach their operating temperature.

When measuring DC voltage, set the function switch to the proper DC polarity, and the range switch to the highest range. Connect the ground lead to the ground or low potential side of the circuit, and then touch the DC probe to the point where the measurement is to be made. Meanwhile, rotate the range switch to the range that gives the greatest deflection without going over full scale, and read the value indicated on the corresponding scale, using the appropriate multiplying factor.

Follow the same procedure for measuring AC voltage, except that you now set the function switch to AC and use the AC-OHMS probe. It may be necessary to adjust the ZERO ADJ control.

In making resistance measurements, you use the same probe as for AC. Before making any measurement, make sure that the power has been turned off in the equipment under test, and that any large capacitors in it have been discharged. Set the function selector to OHMS and the range switch to the nearest value to the resistance being measured. Short the probe and ground leads together, and zero to the meter pointer with the ZERO ADJ control if necessary. Disconnect the leads and adjust the OHMS ADJ to set the pointer to full scale on the ohms scale. Connect the ohms and ground leads on each side of the resistance to be measured. In circuits where polarity is important, remember the probe is *positive*. Reset the range switch if necessary to get a better deflection, and read the value indicated on the ohms scale, multiplying it by the range factor.

We have covered the VTVM in some detail because it serves as a good introduction to a great many electronic instruments, as well as being an important multimeter. Range and function switches are used in nearly all types of test equipment, and the examples you have seen here are typical. In some instruments you will find pushbuttons instead of rotary switches, and in others automatic ranging is done by electronic switching, and function selection by remote control through logic circuits. But these are simply more convenient ways of switching. Range selection still remains a process of tapping down on a voltage divider, or connecting different values of series of parallel resistance, while function selection is the routing of the input signal through the proper circuits to convert it to the form required to operate the indicating device or, as in signal generators, to the form required in the output (as you'll see in the next chapter). In some more elaborate instruments, function selection is performed by the use of plug-in modules, which we shall discuss as we come to them. The principle is always the same, only the methods are different.

Other Electronic Voltmeters

The solid-state equivalent of the VTVM is the TVM, or *transistorized voltmeter*. In principle it resembles the VTVM, but has two features that make it preferable for some applications. These are:

Battery operation (in many cases, both line and operated);
Compactness.

A battery-operated instrument is necessary for field work wherever a VOM would not be sensitive enough. It is also the only way, sometimes, of avoiding troublesome ground loops caused by the power cord of a line-operated instrument (see Chapter 1 for discussion of grounding problems).

Transistor circuits also allow the instrument to be made smaller and lighter, thus making it more portable. There is also no wait while it warms up and stabilizes, and no heat to dissipate.

Electronic voltmeters with amplifiers preceding the meter have greater sensitivity, but are generally designed for special purposes, and cost more. They are, therefore, not usually multimeters, but because they are electronic voltmeters we shall discuss them here before going on to digital voltmeters.

The DC Electronic Voltmeter has either a direct-coupled or chopper-stabilized amplifier. Direct-coupled amplifiers are less expensive, but have a tendency to drift, so that zero adjustments have to be made as in a VTVM. This is not necessary with a chopper, because it changes the DC to AC before applying it to an AC amplifier. After amplification the signal is rectified before it reaches the meter. It seems complicated but, as you saw in Chapter 1, AC amplifiers can be made much more stable than DC amplifiers because of the use of feedback. This is particularly important where the signal needs much amplification, as in DC millivoltmeters and DC null meters.

Some of these instruments also provide for DC current measurement. This is usually done by measuring the small voltage drop across a low-value resistor through which the current to be measured flows. A clip-on milliammeter probe, somewhat like the AC clamp ammeter, but sensitive to DC, is also used.

AC Electronic Voltmeters obviously have AC amplifiers, afterwards converting the AC to DC for application to the meter. There are three categories of AC electronic voltmeters:

Average-responding   Peak-responding
RMS-responding

Most AC voltmeters are average-responding, as you have seen already. Their meters read RMS voltage values, which are greater by a factor of 1.11, but this is true only of sinusoidal waveforms. True RMS measurements are performed by sensing the waveform's heating power, using thermocouples. In this way, the RMS value of any waveform can be measured accurately.

Peak-responding voltmeters use a rectifier in the input circuit to charge a capacitor to the peak value of the voltage being measured. This is then amplified by a DC amplifier and applied to the meter. Peak-responding voltmeters are usually calibrated in the RMS value of a sine wave, and are consequently subject to error if a non-sinusoidal voltage is measured.

AC voltmeters that can be measured RF voltages are called RF voltmeters. The ordinary AC voltmeter's input impedance, amplifier and other circuits limit the frequency to which it can respond because of the losses from distributed capacitance. To overcome this, a diode probe (as illustrated in Chapter 1) is often used. This converts the AC to DC with minimum loss from distributed capacitance. However, for higher frequencies more reliable results can be obtained by using a *sampling voltmeter,* which constructs a low-frequency equivalent of the high-frequency signal. It can then handle this like an ordinary AC voltmeter. RF voltmeters are usually average-responding voltmeters.

DIGITAL MULTIMETERS

The principle of operation of a digital multimeter is quite different from that of a VTVM, at least as far as the DC-voltmeter part of it is concerned. DMMs adjust a precise internal reference voltage to balance the voltage being measured, and convert the time or number of steps taken to do this into a numerical readout.

The simplest way of doing this is by means of a *voltage ramp*. A voltage ramp is produced in the same way as a *sawtooth voltage*, as explained in the chapter on oscilloscopes. It can be rising or falling, but it must be linear. In Figure 6.11, the ramp voltage is shown falling at a steady rate from +12V to −12V. It is applied to a *comparator*, which compares it continuously with the unknown voltage E_x. When it is equal to E_x, the comparator generates a signal that opens a gate to allow pulses from a precision oscillator to pass to electronic counting circuits. A second comparator senses when the ramp voltage reaches zero, and sends a pulse to close the gate. The number of pulses counted depends on the time interval between t_1 and t_2, or how far E_x is above zero.

The general arrangement of a ramp-type DMM is illustrated in Figure 6.12. The input voltage is routed through the *attenuator* (range control) circuit to the input comparator. The *ramp generator* is turned on by the *sample rate control,* and produces the voltage ramp, which is applied simultaneously to the input and zero comparators. When the ramp and input voltages coincide, the input comparator opens the *oscillator* gate, and the *decade counting units* (DCUs) begin counting the pulses. The *zero comparator* closes the gate, so that the *readout* displays as a voltage the total number of pulses counted between t_1 and t_2. In instruments with *automatic ranging*, the DCUs automatically adjust the degree of attenuation to be applied to the input signal.

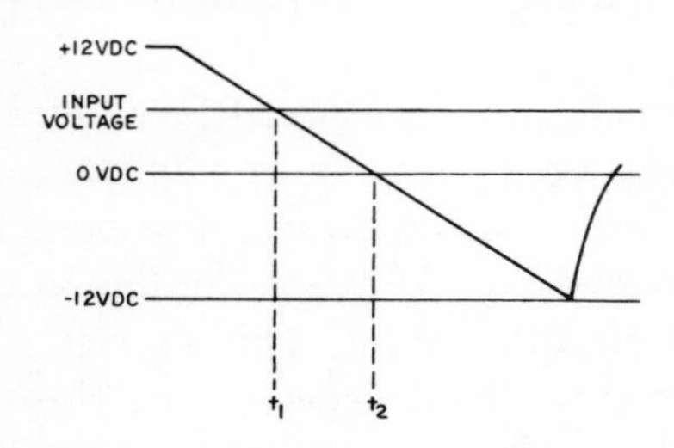

Figure 6.11 How a Ramp-Type Digital Multimeter Works

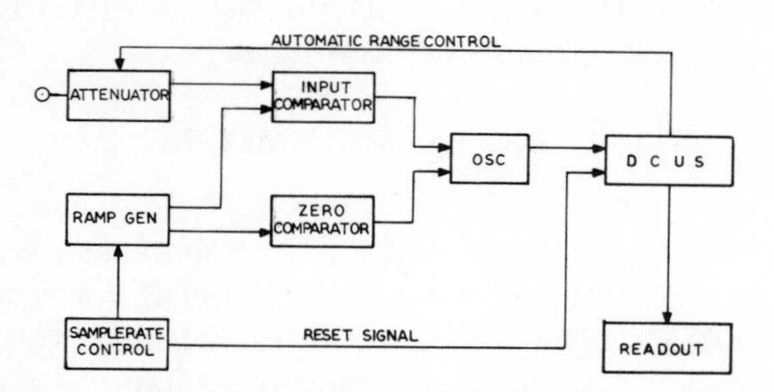

Figure 6.12 Block Diagram of Ramp-Type DMM

The sample rate control is a timing circuit that starts each voltage ramp. The rate is selectable in some DMMs, fixed in others. This circuit simultaneously sends the reset signal to the DCUs to return them all to zero before starting the next count.

Since the ramp voltage does not exceed +12 volts, an attenuator is required to bring higher input voltages within this range, just as was done in the VTVM. This can be done with a manual range switch, or accomplished automatically, as mentioned above. Automatic ranging uses transistor-logic switching in response to signals from the DCUs, introducing attenuation as needed. Both methods control the position of the decimal point in the readout.

The DCUs are integrated circuits consisting of flipflops turning on and off with successive pulses, so that each pulse causes the next higher figure to appear in that position in the readout controlled by the particular DCU. Every tenth pulse is also passed to the next position to the left. This is the electronic equivalent of a mechanical odometer such as that in your automobile (see Chapter 1). Of course it operates at a very much higher speed, because the oscillator is likely to be running at 100 kilohertz.

The readout consists of a row of numeric indicator tubes or solid-state segmented displays. These are each driven by the corresponding DCU, which may be combined with it in one module. The display usually consists of the figures, decimal point and polarity of the measured value, and whether it is DC, AC or ohms.

The figures displayed at time t_2 remain displayed until the next t_1, when they are reset to zero, and the count is repeated. Each count is called a "sample" and several samples are taken every second.

Of course, if the voltage changes, the reading changes too, so that if it is unsteady the reading will jump about. This would happen if there was much noise on the voltage being measured, so this type of DVM does not work so well under noisy conditions.

The ramp-type DMM is comparatively simple, and therefore less expensive than more advanced types. It usually has an accuracy of 0.1 or 0.05 percent of reading, and is capable of measuring AC volts and resistance by converting these inputs to DC, as in the VTVM.

Another type is the *staircase-ramp* DMM, shown in block form in Figure 6.13. The sample-rate control resets the DCUs as before, and it

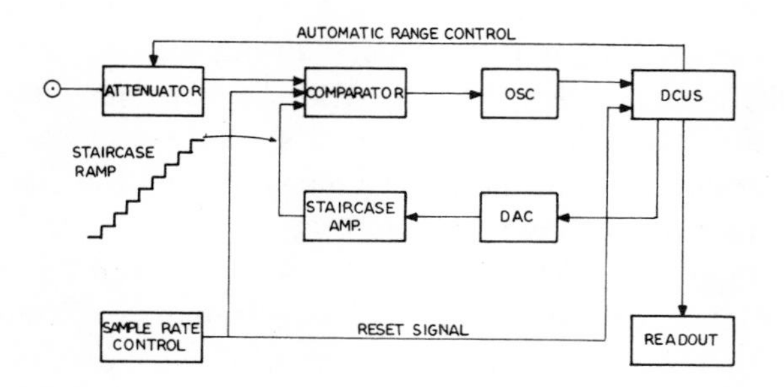

Figure 6.13 Block Diagram of Staircase-Ramp-DMM

gates the oscillator signal to the DCUs. These begin counting, but each counting step not only advances the readout but also is converted into one step of the staircase ramp. This is done in the *digital-analog converter* (often abbreviated D/A converter, or DAC). The amplified ramp is applied to the comparator. When the staircase voltage coincides with the input voltage, the comparator cuts off the oscillator pulses, so that the readout shows the total number of staircase steps as a voltage. Note that the ramp is ascending in this case instead of descending, so no zero comparator is required because the ramp starts from zero.

You can see that in other respects it is very similar to the ramp-type DMM. It has the same disadvantage with respect to noise, however. In some models this is overcome by adding an *integrator* to the circuit, as shown in Figure 6.14. An integrator enables the voltmeter to measure the true average of the input voltage over a fixed measuring period instead of at the end of it.

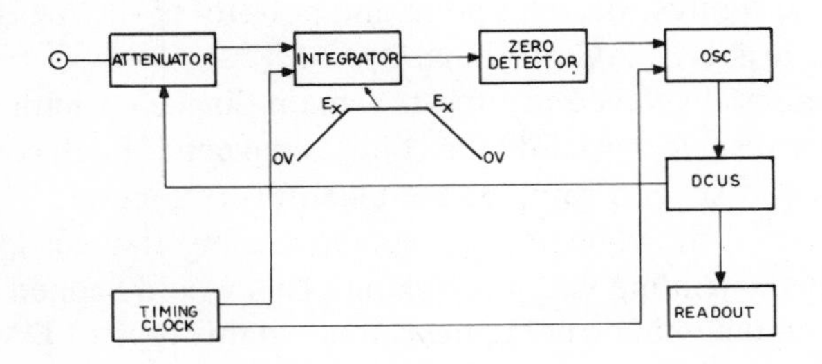

Figure 6.14 Block Diagram of Dual-Slope Integrating DDM

The input voltage is allowed to charge a capacitor for an exact period of time, so that the charge is proportionate to the voltage to be measured. Timing is done by a circuit that counts off the exact number of pulses from the oscillator, and then cuts off the input voltage. In this way, any noise or fluctuation of the voltage sample is ''homogenized'' as it were. Then the capacitor is discharged at a fixed rate, so that it produces a ramp voltage, and the DCUs count the pulses during the time taken to reach zero, as in the ordinary ramp-type DMM. At this point the zero-detect circuit stops the count. The readout now shows a voltage reading that equals the average value of the input voltage during the sample period. One of these DMMs is shown in Figure 6.15.

Another type of DMM, shown in block diagram in Figure 6.16 and illustrated in Figure 6.17, uses a *null-balance* technique. The input circuit converts the unknown voltage to a proportionate value on a 0-10V range. For example, if the unknown voltage is 50 volts, and the DMM range selected is 0-100 volts, the converted voltage will be 5 volts. When the display (sample) rate circuit switches the stop-start circuit to start, the DCUs start counting clock pulses. The decades count in sequence, beginning with the 10^3 decade. At each count a resistor allows a certain value of current to flow from the reference supply to the null detector. With each succeeding count another resistor is switched in parallel with the

Figure 6.15 Dual-slope integrating DMM with pushbutton controls has 74 dB noise rejection and 0.005% DC accuracy. *(Courtesy Dana Laboratories, Inc.)*

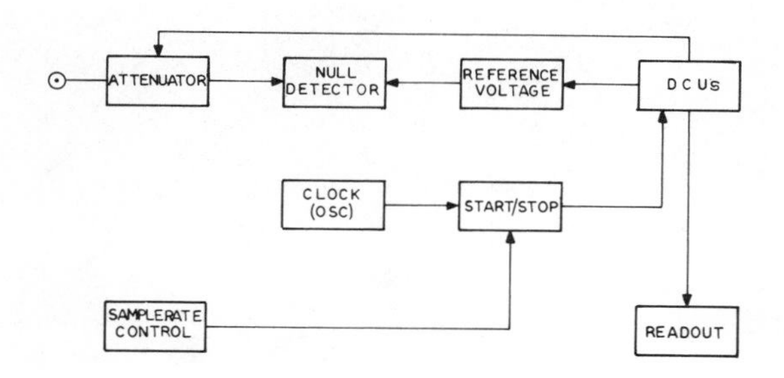

Figure 6.16 Block Diagram of Null-Balance DMM

first, increasing the reference current until the null detector senses it has become equal to, or greater than, the input current. The 10^3 decade stops counting at the last figure displayed, and the 10^2 decade repeats the performance, followed, in turn, by the 10^1 and 10^0 decades. In this way, the value displayed approaches the actual value of the input by successive approximations. This process, which is called "digitizing," takes less than two milliseconds to complete.

Like other multimeters, DMMs measure DC voltage. Front-end circuits are required to change AC or resistance into DC, as in the VTVM. In some DMMs these are built in, in others they are supplied by interchangeable plug-in modules. The desired function is selected by a function control, as in a VTVM.

DMMs are used in preference to other types in automated systems of test equipment because their data is in digital form, so can be fed directly into a computer or a printer. Remote control by computer is also provided, in many cases, by additional logic circuits in the DMM. In many manufacturing processes, the product (a missile guidance system, for example) is subjected to a comprehensive functional test lasting several days, in which its entire performance is appraised by a computer. This computer also controls the test equipment, and if a voltage measurement is required, it selects function and range by remote control, and connects the DMM to the test point. It then compares the reading with data stored in its memory, and causes a printer to type out the actual value, together with the high and low permissible limits. A measurement within these limits is "GO," one outside them is "NO-GO." In the latter case, the product has failed the test, and may have to be repaired.

Figure 6.17 Null-Balance (Successive Approximation) DMM has 80 dB noise rejection and 0.01% DC accuracy. *(Courtesy Dana Laboratories, Inc.)*

On a smaller scale, there are available DMMs that determine by themselves whether the measurement is valid. This type has dials whereby the limits may be preset. Then if the measurement is within the limits, a green light lights, if outside, a red or yellow light comes on. The operator has only to make the connections, and respond to the lamp indication. The DMM uses a voltage ramp, with comparators for each limit.

DIFFERENTIAL VOLTMETERS

In a Wheatstone bridge, you measure an unknown resistance by adjusting a standard resistance until the potential drop across it equals the potential drop across the unknown. You know when they are equal because the null meter reads zero. The value of the unknown resistance is given by the figures on the decade dials used to adjust the standard resistance, as you saw in Chapter 3.

In a differential voltmeter (DVM) you measure an unknown voltage by adjusting a precision voltage divider until the potential drop across it equals the unknown voltage, as indicated by a null meter. The value of the unknown voltage is given by the figures on the decade dials of the precision voltage divider, as you can see in Figure 6.18.

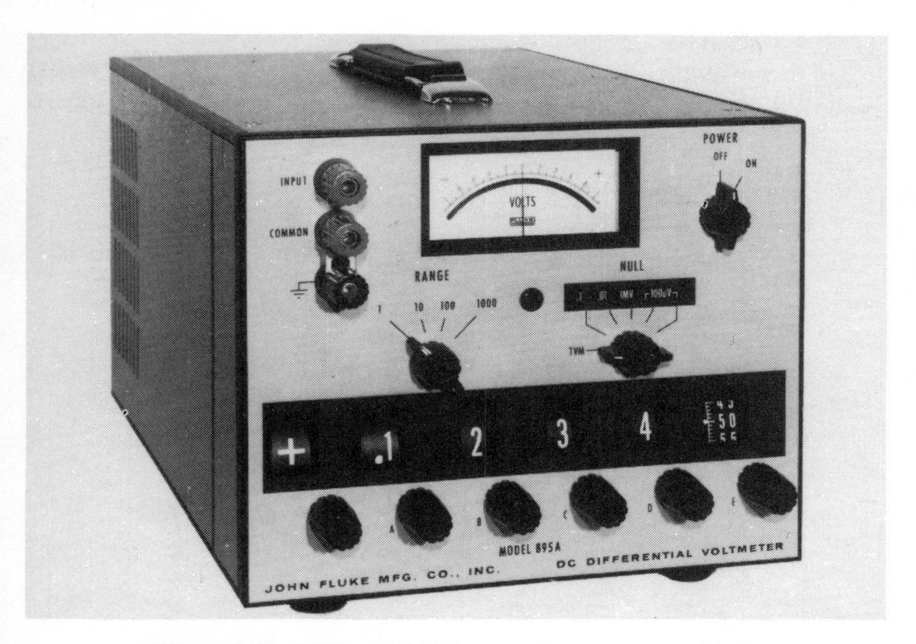

Figure 6.18 Differential Voltmeter *(Courtesy John Fluke Mfg. Co., Inc.)*

The difference between this instrument and a Wheatstone bridge is that in the resistance bridge you adjust a standard resistance until it balances the unknown resistance, whereas in the DVM you adjust a standard voltage until it balances the unknown voltage. The laboratory method of doing this is to use a precision potentiometer as a voltage divider, referenced to a standard cell, and used in conjunction with a galvanometer, working battery, and other items, that are time-consuming to hook up, and require skill and experience to operate properly.

The differential voltmeter is a streamlined potentiometric system with everything built in, that is altogether simpler and safer to use.

Instead of a vulnerable standard cell this instrument uses specially-selected zener diodes to hold the voltage to the reference value regardless of line-voltage variations, temperature, component aging or other variables. The model illustrated succeeds in doing this so well that the total drift of its reference voltage *in the course of a year* is less than 20 parts per million (0.002%).

To take full advantage of this accuracy, the voltage divider must have comparable precision. This is accomplished by using a Kelvin-Varley voltage divider. As you can see from Figure 6.19, this is a set of four decades of precision wirewound resistors, plus a high-resolution potentiometer.

In this example, the reference voltage is 100V, and is applied across decade A. Each of the 11 resistors in this decade has a value of 10

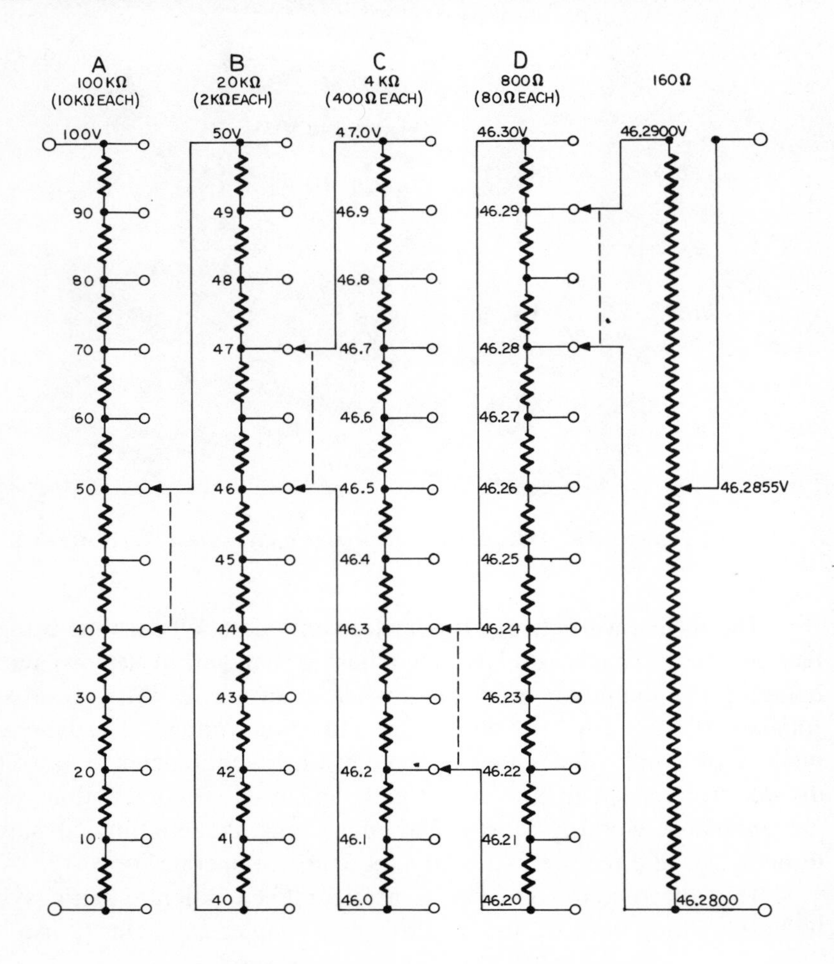

Figure 6.19 Kelvin-Varley Voltage Divider

kilohms. When the decade knob is rotated, the double slider moves up or down, connecting the pair of resistors between the contacts to the opposite ends of decade B. As a result, the two A resistors (20 kilohms) are shunted by the 20 kilohms of decade B, so that their value is actually only 10 kilohms. That is how a decade of eleven 10-kilohm resistors manages to have a total resistance of 100 kilohms instead of 110 kilohms!

The two sliders on the A decade connect the voltages at their points of contact to the opposite ends of decade B, consequently the voltage at the top of B is 50V (in this case), and at the bottom it is 40V. B's sliders are shown connecting 47V to the top of decade C and 46V to the bottom. The pair of 2-kilohm resistors between B's sliders also total only 2 kilohms, because they are shunted by the whole of C, which has a value of 4 kilohms.

In the same way, the C and D decades subdivide the voltage further, so that the decade dials show its value to four figures. Two more figures are given by the high-resolution potentiometer.

In this way, the Kelvin-Varley voltage divider matches the reference voltage exactly to the voltage being measured. The principal circuits in the DVM are shown in Figure 6.20 in block form. The voltage to be measured is amplified in the DC amplifier, which is also a combination of VTVM and null detector. In the example, the voltage (46.2855V) would have been nulled with the RANGE switch in the *100* position. For other voltages this switch would be set accordingly. The decimal point automatically positions itself as the RANGE switch is set.

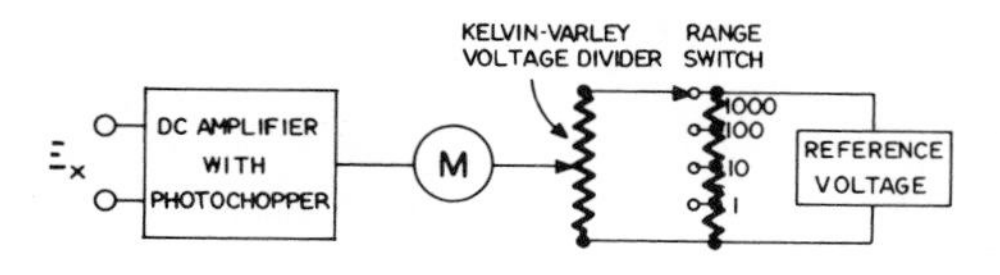

Figure 6.20 Block Diagram of DVM

The instrument illustrated in Figure 6.18 is therefore very simple to use in spite of its great accuracy (±0.0025 percent under laboratory conditions). To measure any potential up to 1000 volts you would do the following:

1. Plug the power cord into a service outlet and turn the POWER switch to ON.
2. Set the RANGE switch to 1000 and the NULL switch to TVM. This connects the instrument as a transistorized voltmeter, to enable you to get an approximate reading of the voltage to be measured. (Read the value on the null meter dial, which also has a voltage scale.) In this mode the input impedance is 100 megohms.

3. Connect the voltage to be measured to the INPUT binding posts. The meter pointer will deflect. If it goes to the left, reset the polarity indication (bottom left) to −. It will now read on the right-hand half of the scale.
4. Switch the RANGE control downwards for maximum deflection of the meter pointer without going off-scale.
5. Turn the NULL switch to *1*, and adjust the decade dials to zero the meter. Repeat this for each position of the NULL switch until the meter has been zeroed at the most sensitive position.
6. Read the value of the unknown voltage from the figures on the decade dials.
7. Return the NULL switch to TVM before disconnecting the voltage.

The polarity switch can also be set to RATIO. This connects the Kelvin-Varley voltage divider across two binding posts in the rear of the instrument. This allows two voltages to be connected at the same time to the DVM, one in front and one behind. When the meter has been nulled, the decade dials indicate the ratio between the two voltages. Output terminals for a recorder are also located on the rear.

The instruments in this chapter have one feature in common. They are all fundamentally DC voltmeters. They may also be used for measuring AC and resistance by converting these functions to DC, but giving the measured value in the appropriate units.

7

HOW TO USE SIGNAL SOURCES FOR TROUBLESHOOTING

You can best judge the performance of an amplifier or similar piece of electronic gear by what it does to a test signal. This usually consists of feeding a suitable signal to its input and viewing the output signal with an oscilloscope. Test signals include *continuous waves, pulses* and *noise*.

Continuous waves are those that continue their periodic excursions without interruption. Typical continuous waves shown in Figure 7.1 are:

Sine waves	Sawtooth waves
Triangle waves	Square waves

Pulses are individual excursions from a baseline with intervals between them called *baseline dwell times*. Typical pulses shown in Figure 7.2 are:

Trapezoidal	Spiked
Rectangular	Rounded

Noise signals consist of random impulses distributed over a frequency band. Noise spectra produced by random-noise generators are:

White	Pink	ANSI

SIGNAL SOURCES

The sources for these signals are:

Oscillators and Function Generators	Pulse Generators
Signal and Sweep Generators	Noise Generators

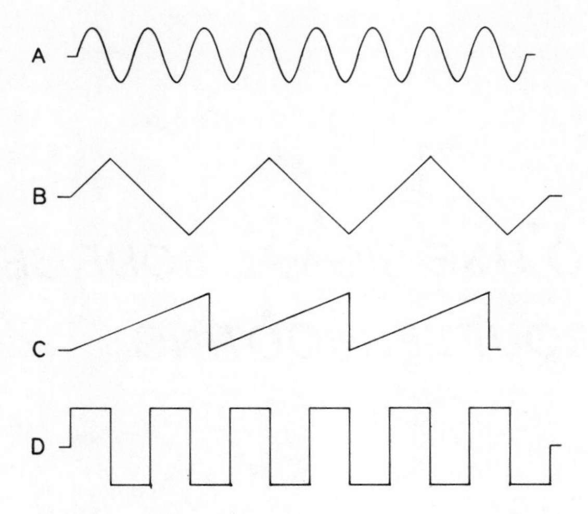

Figure 7.1 Continuous waves have no intervals between their excursions. A = Sine Waves; B = Triangle Waves; C = Sawtooth Waves; D = Square Waves.

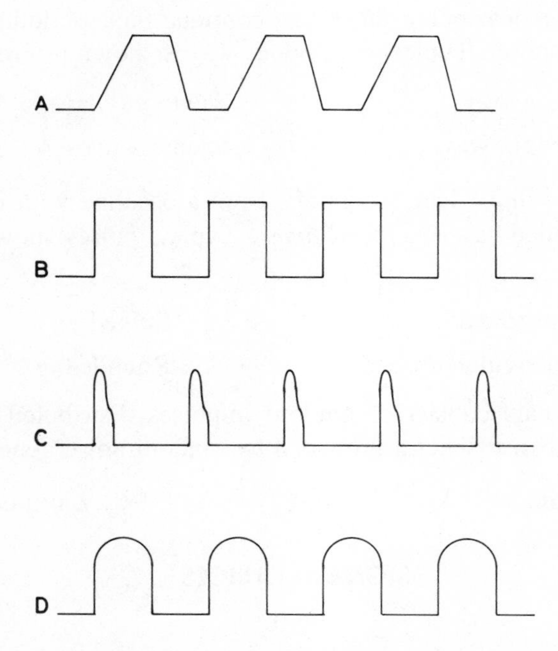

Figure 7.2 Pulses have intervals between their excursions called baseline dwell times. A = Trapezoidal Pulses; B = Rectangular Pulses; C = Spiked Pulses; D = Rounded Pulses. (Pulses are idealized.)

The *oscillator* generates sine-wave signals of known frequency and amplitude. If it is called an *audio oscillator,* it has a frequency range from 20 hertz to 20 kilohertz or higher. An *audio-signal generator* has a similar range, but higher power. A *test oscillator* is a solid-state oscillator with a greater frequency range. The last two usually have meters to show the amplitude of the output signal.

Function generators are oscillators that produce triangle, sawtooth or square waves in addition to sine waves. (There are also generators that produce only one each of these functions.) They operate at frequencies from 100 kilohertz down to less than one hertz.

Signal generators are oscillators that produce higher frequency signals that can be modulated. These include RF signal generators (50 kilohertz through 40 gigahertz) with AM, FM or pulse modulation capability.

Sweep generators are an extreme case of frequency modulation. They put out a signal that continuously increases its frequency at a steady rate, starting at a lower limit and sweeping up to an upper limit, from which it returns immediately to the lower limit to start the next sweep. These are usually adjustable. Most have frequency markers. We shall discuss sweep generators again in this chapter, and later on in the chapter on TV and radio test equipment.

Pulse generators produce high-quality rectangular test pulses, with repetition rates mostly not exceeding one megahertz. The most important characteristics in a pulse generator are the *rise time* and *shape* of the pulse. We shall come back to these later on in this chapter.

Noise generators generate random electrical noise from carbon-resistor, semiconductor, gas-tube or other source, over a bandwidth from 2 hertz to 20 megahertz (depending on the instrument).

IMPORTANT FEATURES IN OSCILLATORS

The most important features of oscillators are their:

Frequency range	Dial resolution
Output voltage or power	Harmonic distortion
Stability	Hum and noise

The *frequency range* of an oscillator is the first feature that determines whether it will be any use to you for the purpose you have in mind. Obviously, you want one that will cover the frequency band you're interested in, with a reasonable margin on either side.

The *output voltage or power* of an oscillator depends upon the type of oscillator and the load placed upon it. Most oscillators are designed to give a certain output with a particular output impedance, such as 600, 75 or 50 ohms. There also are some in which the impedance may be varied.

Stability is important for both frequency and amplitude. Frequency stability is obtained by using large amounts of negative feedback, precision components and carefully regulated voltages. Since temperature also affects frequency, an oscillator must be allowed an adequate time to warm up to its operating temperature when first turned on, after which it should not change appreciably. The amplitude should not alter as the frequency is varied, but remain constant over the entire range of the oscillator.

Dial resolution is a measure of how close the oscillator's frequency at its output terminals is to the frequency indicated on its tuning dial.

Harmonic distortion occurs because harmonics of the fundamental frequency are also generated, and can degrade the purity of its waveform. Some harmonic distortion is always present, but as long as it is only some fraction of one percent it is negligible.

Hum and noise must be kept as low as possible for the same reason. As we saw in Chapter 1, this requires careful shielding, grounding and filtering.

OSCILLATOR CIRCUITS

As you know, there are LC and RC oscillators. Each type has features that make one or the other preferable for this or that application. Modern test oscillators favor RC circuits because they are less cumbersome, more stable and not as susceptible to EMI and AC pickup. For sine-wave oscillators, the Wien-bridge and bridged-T circuits are often used, while multivibrator circuits are employed in function and pulse generators.

Wien-Bridge Oscillator

A typical Wien-bridge oscillator is illustrated in Figure 7.3, and its principle of operation is diagrammed in Figure 7.4. The requirements of the bridge are that R1 = R2 and C1 = C2, and that (R3 + R4) = (2 x RT1). When this is so the bridge will be resonant at the frequency given by:

$$f = \frac{1}{2\pi R_1 C_2}$$

C1 and C2 are two sections of a split-stator variable capacitor, so that C1 will equal C2 at all settings. The rotor of this capacitor is turned by the frequency control knob. R1 and R2 are really pairs of identical resistors mounted on the range switch, so that at each position of this control (X1, X10, X100, X1000) they will be both equal and able to provide the proper multiplying factor for the selected range. If the frequency dial is marked with graduations from 5 to 60, say, then when the range setting is X1, the output frequency will be from 5 to 60 hertz, according to the dial position; if the range setting is changed to X10, the output frequency will be variable from 50 to 600 hertz; and similarly for the other two ranges.

Oscillation is sustained by positive feedback paths from the cathodes of V3 and V4. Should the amplitude of this signal, which is also the output signal, tend to increase or decrease, the signal current through the bridge will also tend to change. In the right-hand side of the bridge is the ballast lamp RT1. Its resistance increases as the current increases, and vice versa, so that it regulates the amplitude of the signals applied to V1

Figure 7.3 This typical Wien-bridge oscillator has a frequency range from 5 Hz to 600 Hz. Notice the shorting trap grounding one side of the output, making it unbalanced. *(Courtesy Hewlett-Packard Co.)*

and V2, to maintain a constant output amplitude. This is further regulated by negative feedback from the plate of V3 to the control and screen grids of V4, and from the plate of V4 to the control and screen grids of V3.

An amplitude control will normally be located at the output, though not shown in this simplified schematic. It will be a variable T-pad (as in Figure 7.5) or similar configuration to maintain a constant output impedance at all settings.

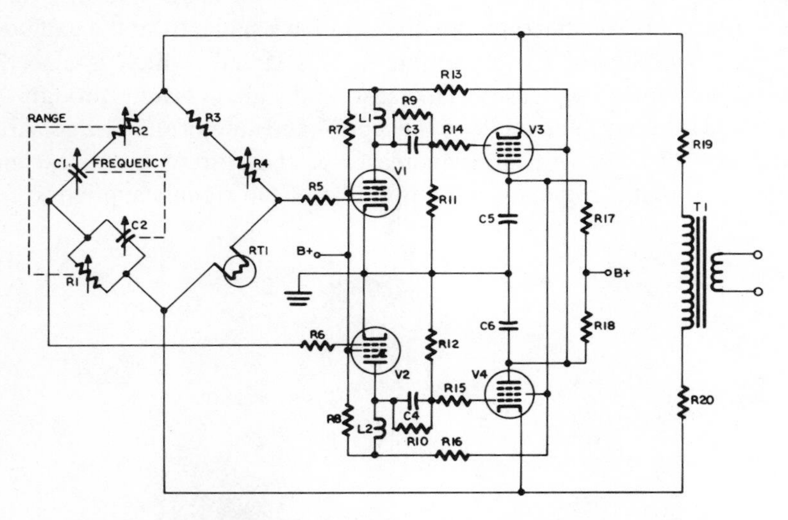

Figure 7.4 Simplified Circuit of Wien-bridge Oscillator (omitting power supply and output attenuator).

Bridged-T Oscillator

The Bridged-T Oscillator is not unlike the Wien-bridge Oscillator, as you can see in the simplified schematic in Figure 7.5. V1 and V2 are a differential amplifier. Positive feedback is applied to V1's grid through RT1, which is a ballast lamp regulating the amplitude of the output signal. Negative feedback reaches V2's grid through the frequency-determining bridged-T network C1, C2, R1 and R3.

Component values are such that the positive feedback exceeds the negative, therefore oscillation is sustained. As the amplitude of the positive feedback increases, the resistance of RT1 increases also, reducing its voltage until it stabilizes a little above the value of the negative feedback.

In the Wien-bridge oscillator (Figure 7.4), C2 equals C1, but in the bridged-T oscillator, C2 = 4 x C1. This ratio causes the harmonics in the feedback signal to be greatly increased. The second harmonic is doubled, the third trebled, and so on. At first this seems to be crazy, since we want to reduce them. However, as this is *negative* feedback, the effect is just what we want—in fact, better! The unwanted harmonics are attenuated to a greater extent than in the Wien-bridge oscillator, resulting in an even greater purity in the waveform of the output signal.

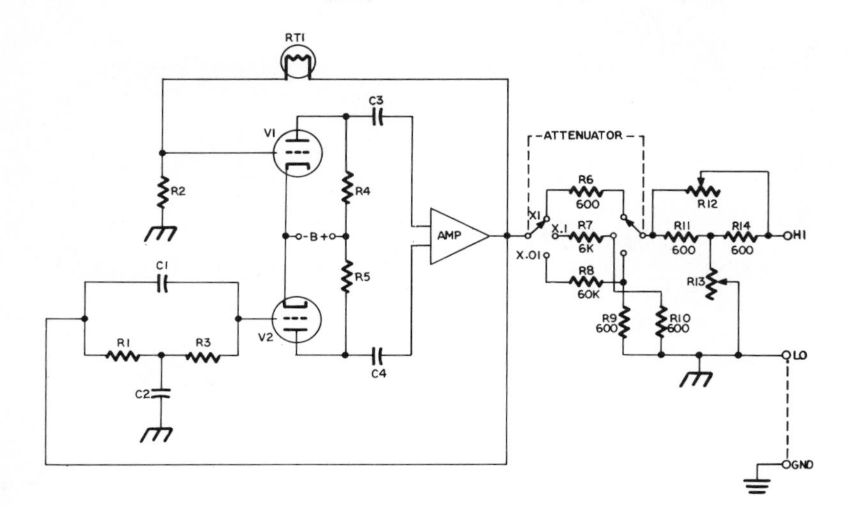

Figure 7.5 Simplified Circuit of bridged-T oscillator (with detail of output step attenuator and bridged-T variable attenuator).

In the output of the bridged-T oscillator shown in Figure 7.5, is a step-attenuator R6 through R10, with three steps: X1, X.1 and X.01. This divides the output voltage by 1, 10 or 100, while maintaining an output impedance of 600 ohms. The vernier bridged-T control (R11 through R14) allows for variable adjustment of the output voltage between the attenuator steps, while also maintaining the output impedance at 600 ohms. It is worthwhile making a special note of these circuits, for they are used quite a lot. Other output impedance values can be provided by suitable choice of components.

Multivibrator Oscillator

A typical function generator, as illustrated in Figure 7.6, is shown in block form in Figure 7.7. A flip-flop generating a square wave output sends this signal to the upper terminal of the function selector switch, where it can be connected to the output amplifier, and thence to the output jack.

At the same time, the square wave is also applied to the *integrator,* in which it charges a capacitor via a resistor to produce a linear triangle wave. This wave rises in a positive direction, with the positive excursion of the square wave, and falls on the negative excursion.

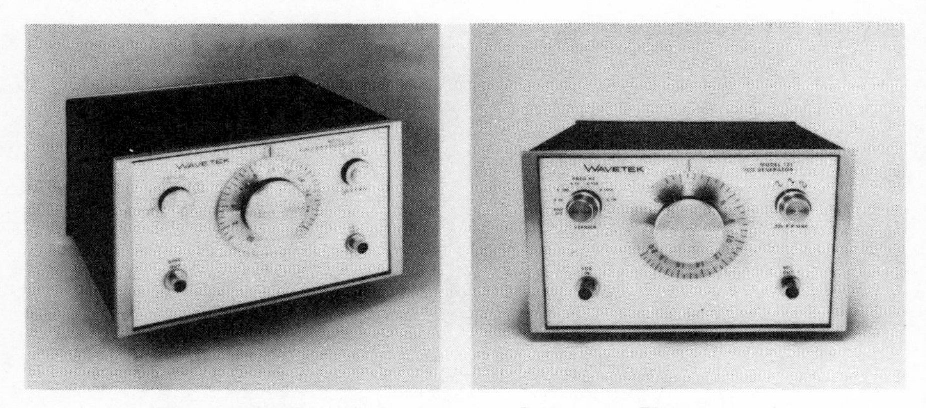

Figure 7.6 Two Similar Function Generators. The one on the right may be voltage-controlled. *(Courtesy Wavetek, Inc.)*

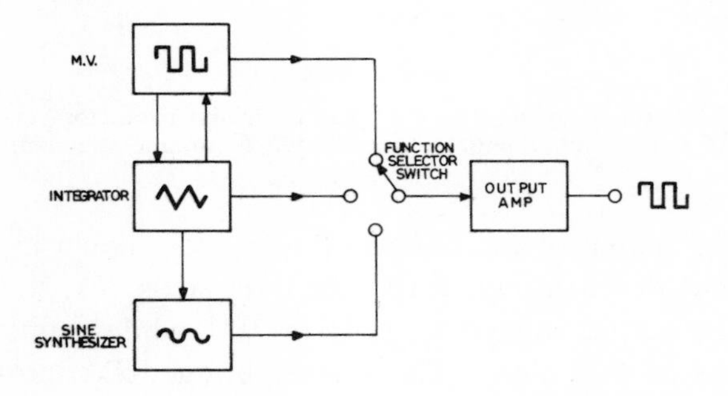

Figure 7.7 Block Diagram of Function Generator (showing multivibrator, integrator, and sine synthesizer).

The frequency of the triangle wave is controlled by selection of the resistance and capacitance values. When the voltage on the capacitor reaches a certain value, it causes the flip-flop to change its state, so that the square wave goes from positive to negative, and vice versa. The time taken to reach this voltage depends on the RC constant of the selected resistor and capacitor. The frequency of both square wave and triangle wave are the same, of course.

The triangle wave is also fed to the *sine synthesizer*. As you can see in Figure 7.8A, this consists of a resistor R1 and a series of shunt resistors (R2, R3, R4, and R5) with a double row of diode switches, biased by a voltage divider.

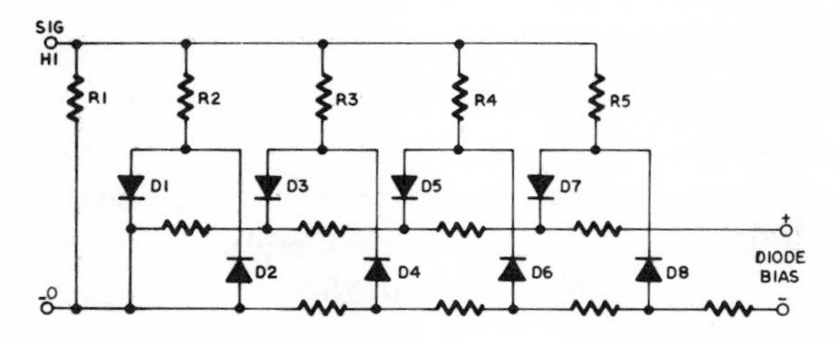

Figure 7.8a Detail of Sine Synthesizer

The triangle wave is applied across R1. As its voltage rises, the voltage across D1, D3, D5, etc., rises also. All the diodes are reverse-biased at first. When the voltage at D1 reaches the level where it becomes forward-biased, R2 is switched into the circuit parallel with R1. The resistance of R1 and R2 in parallel is less than R1 alone, so the voltage across R1 now rises at a slower rate. As the voltage rises further, diodes D3, D5 and D7 switch on in turn, adding R3, R4 and R5 in parallel with R1 and R2. Each resistor in turn decreases the slope of the triangle wave further.

When the triangle wave reaches its apex, its slope reverses and the voltage across R1 declines. One by one, the diodes become reverse-biased again and switch off, removing their shunt resistors from the circuit until the original slope angle of the triangle wave is restored, but in the reverse direction. The slope now continues into the negative region, where the switching is done by the second row of diodes, with opposite polarity. On reaching its negative peak, the triangle wave changes direction again, and these diodes switch off sequentially in the same way as the first row on the positive excursion.

The effect on the triangle wave is shown in Figure 7.8B, where you can see that it now has the approximate shape of a sine wave. Although not perfect, it is within one percent of a pure sine wave, which is close enough for all practical purposes.

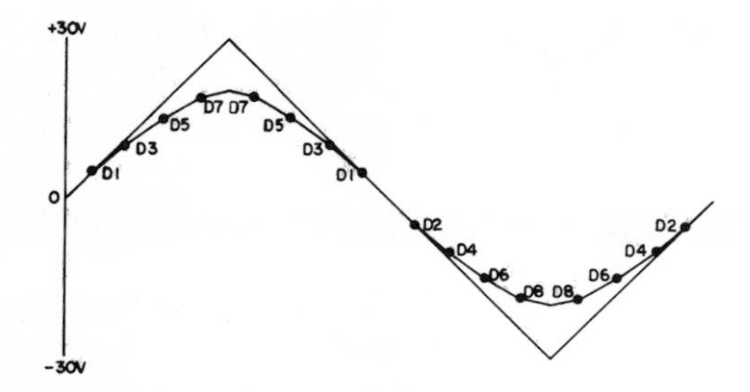

Figure 7.8b How the Sine Synthesizer (A) changes a triangle wave into a sine wave

A *voltage-controlled generator* (VCG) is also illustrated in Figure 7.6. This instrument is almost identical to the function generator. However, VCGs use voltage-variable capacitors (VVCs) in their frequency-determining circuits. When you vary the voltage across a VVC, you change its capacitance, so you can remotely control the frequency of a VCG by varying the input voltage to the VCG IN connector instead of using the tuning dial.

SIGNAL GENERATORS

Signal generator circuits are very similar in principle to radio-transmitter circuits. Basically they consist of an *oscillator* and a *power amplifier*. A *buffer amplifier* is often provided between these to give isolation, so that load variations do not affect oscillator frequency. Figure 7.9A illustrates a typical AM signal generator, and Figure 7.9C shows a more elaborate one.

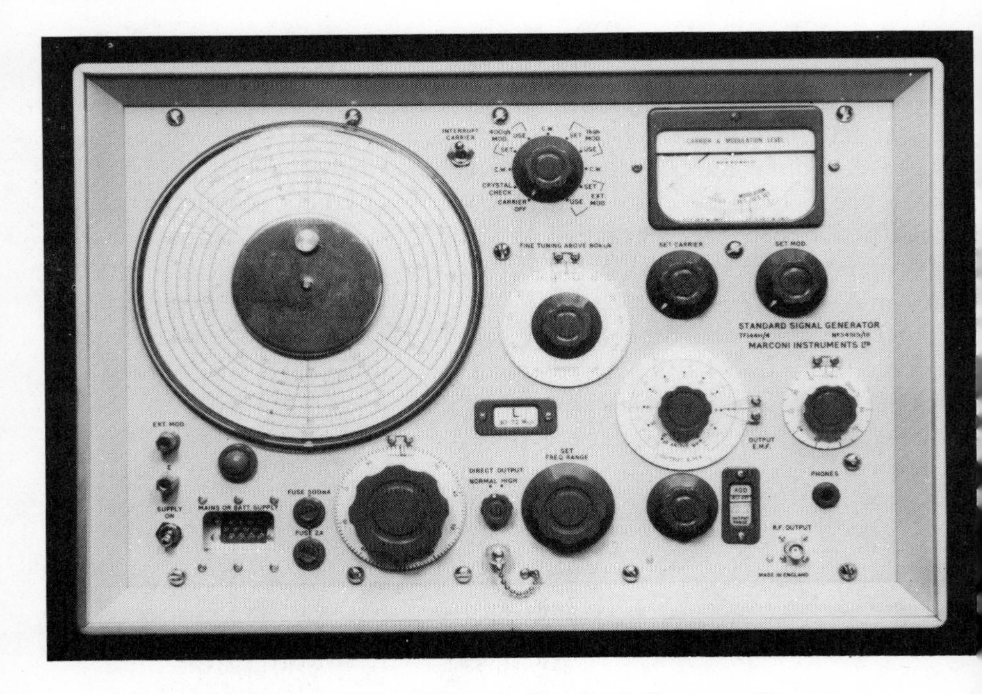

Figure 7.9a Typical good-quality Signal Generator *(Courtesy Marconi Instruments)*. For exaplantion of controls, see Figure 7.9b.

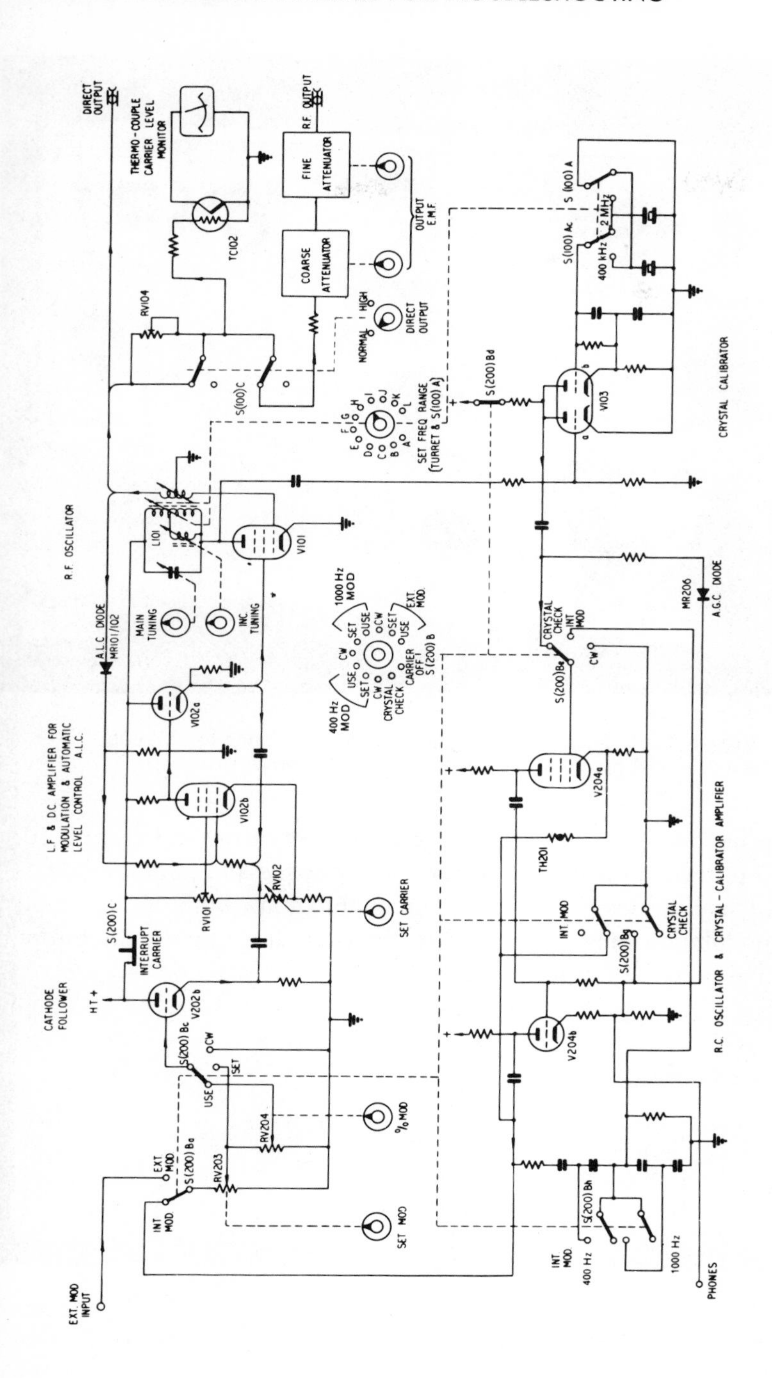

Figure 7.9b Functional Diagram of Signal Generator in (A)

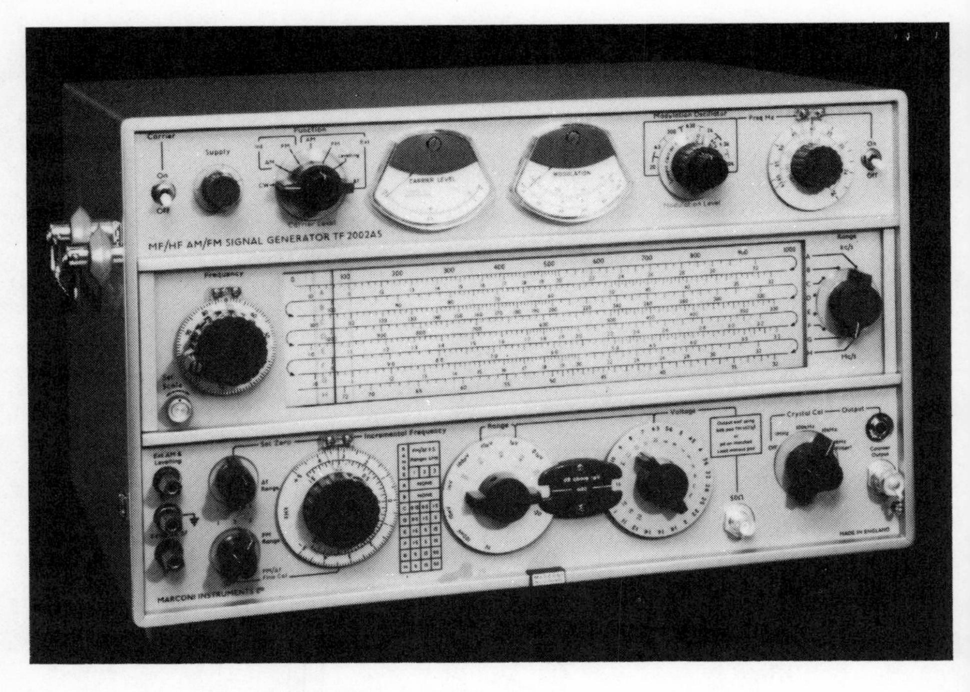

Figure 7.9c This AM/FM signal Generator is similar to (A), but with the FM function as well. *(Courtesy Marconi Instruments)*

The principal circuits of a signal generator are shown in Figure 7.10. In addition to those already mentioned, you will see an internal modulation circuit and provision for external modulation. Both internal and external modulation may be applied to the oscillator circuit, the buffer circuit or the power amplifier.

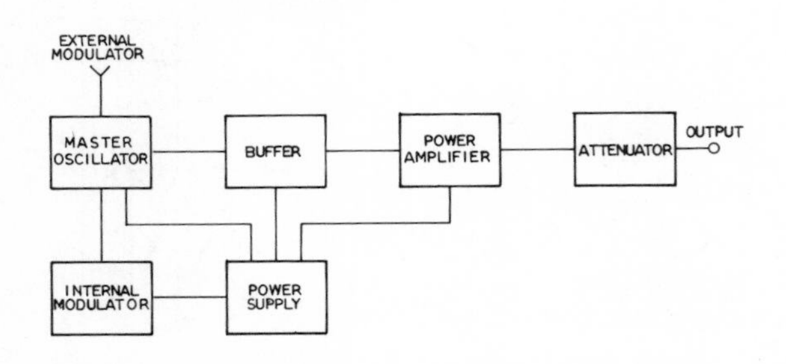

Figure 7.10 Block Diagram of Signal Generator

For low frequencies, as in an audio signal generator, the oscillator will be similar to the RC types we have already discussed, but at higher frequencies LC circuits are used, with one or two tubes or transistors operated Class C. Tuning is done by a variable capacitor. In instruments for use at microwave frequencies, a klystron with variable cavity is used as an oscillator.

Because of its higher frequency, an RF signal generator must be better shielded to reduce stray coupling, and have better frequency stabilization. A variation of one precent at 1 kilohertz is only 10 hertz, but at 10 megahertz it's 100 kilohertz. Increased frequency stability is obtained by the use of a buffer amplifier, as already mentioned, and by a regulated power supply. The principal source of frequency instability, however, is due to temperature changes. A signal generator should, therefore, be given time to reach its normal continuous operating temperature, and the ambient temperature of its environment should be reasonably constant.

A good signal generator needs a good attenuator in its output stage. The T and bridged-T types are used, sometimes with several steps, as in Figure 7.11. Each step is carefully shielded because in a high degree of attenuation complete isolation of the output from the input of the attenuator is essential.

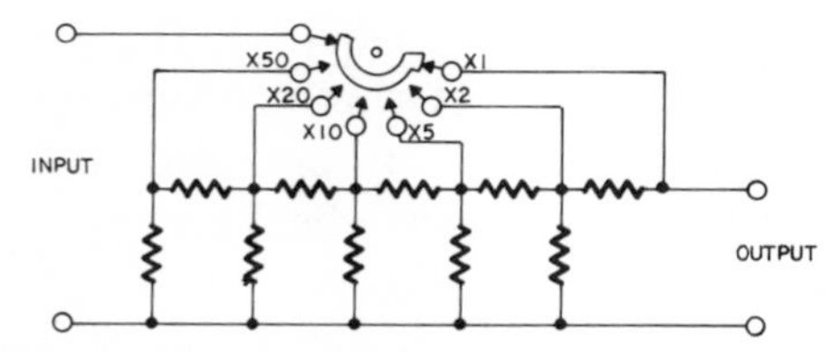

Figure 7.11 Step attenuators like this one are used in signal generators. Because of the higher frequencies, and also because the attenuator factor may be as high as 10,000 to 1, each step must be heavily shielded, so that the input is isolated from the output.

The attenuator must be properly terminated if its calibration is going to mean anything. The output impedance of the attenuator should be the same as the load impedance, unless the load impedance is very much higher. On many signal generators an output meter is provided to show the output in volts and decibels. The readings of the meter will be correct only with the proper load.

The maximum transfer of power from the generator to the load takes place when the output impedance of the generator equals the input impedance of the load. If they do not, there is an impedance mismatch. When connecting two dissimilar impedances, you avoid a mismatch by using a *matching pad*.

A matching pad is a two-way "mask" that makes the load impedance look like the generator impedance to the generator, while making the generator impedance look like the load impedance to the load. Each ''sees'' an impedance equal to its own.

For instance, suppose the sweep generator in Figure 7.13 has an output impedance of 50 ohms, while the input impedance of the unit under test is 300 ohms, and you have a coaxial cable with a characteristic impedance of 50 ohms to make the connection. You'll need a matching pad at the input of the unit under test. This pad should be a *minimum-loss pad* for minimal attenuation of the signal, unless you also want attenuation. It can be constructed with two resistors, as shown in the inset in Figure 7.13. The generator sees R1 shunted by R2 in series with Z2. The total resistance of this combination comes to 51 ohms. The unit under test sees R2 in series with R1 in parallel with Z1, amounting to 296 ohms. Both values are close enough for all practical purposes.

You can make pads like this or you can buy ready-made ones. Some test equipment manufacturers have available a great many different types for all impedance combinations and with various degrees of attenuation.

Modulation signals are generated internally by a separate oscillator built in the signal generator or applied externally. The internal signals are usually fixed-frequency ones of 400 or 1000 hertz, external modulation being used if other frequencies are required. The percentage of modulation can be set, and read on a meter on the front panel.

Many RF signals generators are provided with an internal crystal calibrator, which puts out frequency-check signals one megahertz apart over the band. A telephone jack allows you to listen with headphones to the beat signal as the variable RF signal approaches the calibrator signal, so that you can check the accuracy of the dial setting, and adjust it if necessary.

SWEEP GENERATORS

Sweep generators are used with an oscilloscope to study the frequency response of the device under test. The oscillator frequency of the sweep generator is made to vary by a modulating signal, so a sweep generator is really an FM signal generator. This may be done by passing the modulating signal through a coil coupled inductively to the main oscillator coil in such a way as to change its inductance. Thus the frequency of oscillation is varied at the rate of the modulating frequency.

The modulating frequency may be fixed or variable. The simpler units use a 60-hertz signal from the power line. Variable sweeps may, however, be obtained from 10 microseconds to 100 seconds duration.

This is called the *sweep time*. The range of change of frequency is the *sweep width*. This may be between an upper and lower unit, or about a center frequency. A typical sweep generator is illustrated in Figure 7.12.

Figure 7.12 This sweep generator doubles as a function generator, and may also be voltage controlled. *(Courtesy Wavetek, Inc.)*

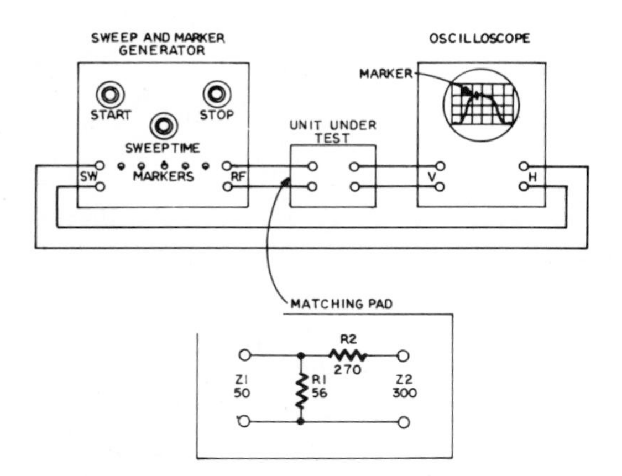

Figure 7.13 Using a sweep and marker generator with an oscilloscope to check the frequency response of the unit under test. A matching pad will be required between dissimilar impedances.

The modulated output of the sweep generator is applied to the input of the device to be tested, and the output of the latter is fed to the vertical input of the oscilloscope. At the same time, a portion of the modulating signal is fed to the horizontal input of the oscilloscope. Each frequency excursion of the sweep generator results in one sweep across the oscilloscope screen. The amplitude of the oscilloscope display corresponds with the response of the device under test to the range of frequencies applied to it (see Figure 7.13). As the sweeps follow each other rapidly, a continuous display remains on the screen. This allows you to "tweak" the tuning points while you observe the effect of each adjustment.

Many sweep generators have single-frequency signal generators built in, by which "markers" may be superimposed on the oscilloscope display to indicate the exact frequency at various points. An instrument designed as a separate source of markers is called a *marker generator*.

PULSE GENERATORS

Pulse Characteristics

Pulses belong in the *time domain,* as opposed to sinusoidal waves, which occupy the *frequency domain*. In both domains the amplitude of the signal is an important parameter. But in the time domain the other parameters are mostly concerned with time intervals, as we shall see below.

Pulse signals have many uses. For example they are used in:

Radar and navigational equipment	Computers
Communications (pulse modulation)	Process control
Telemetry	

Pulses have various shapes. As mentioned at the beginning of this chapter, they are distinguished from continuous waves by having intervals between their individual excursions.

The *baseline* (OT in Figure 7.14) is the reference line for a pulse. It may be at zero potential, or at a positive or negative potential. In the latter case, it is said to be *offset* by the value of the DC potential. Dimensions normal to the baseline represent amplitude, those parallel to it repesent time. The time during which any part of the pulse waveform coincides with the baseline (zero amplitude) is the *baseline dwell time*. Pulses may make excursions on either side of the baseline, but must have baseline dwell times between excursions.

A *trapezoidal pulse* has a flat top parallel to the baseline, with a leading edge that slopes up from the baseline to the top, and a trailing edge that slopes downward from the top to the baseline, as in Figure 7.2.

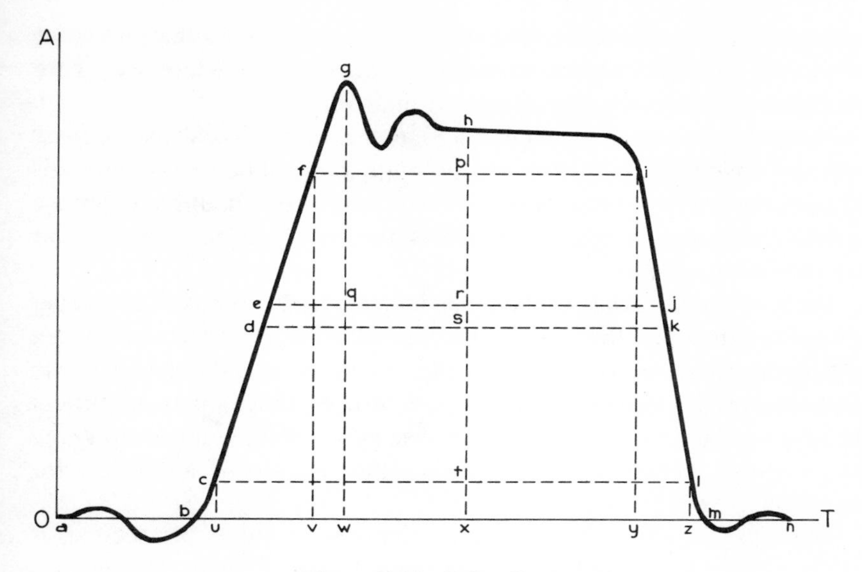

Figure 7.14 Pulse Geometry

gw = peak amplitude

e = leading edge peak median

j = trailing edge peak median

h = centrum: er = rj

hx = pulse amplitude: hs = sx

d = leading edge median

k = trailing edge median

dk = pulse width

c = leading edge decium ("10% point"): cu = 10% of hx

f = leading edge nonum ("90% point"): fv = 90% of hx

uv = risetime

i = trailing edge nonum

l = trailing edge decium

yz = fall time

abcd = initial corner

defgh = leading edge corner

hijk = trailing edge corner

klmn = terminal corner

As the leading and trailing edges are made steeper, the pulse approaches the shape of a rectangle. It can never become a perfect rectangle because to do so the amplitude would have to rise from the baseline to the top in zero time, which is not possible. Some pulses look on an oscilloscope screen as if they do just that, but if the sweep speed is increased you can see it is only because of the terrific speed of their rise and fall that such pulses *seem* to be rectangular.

Rectangular pulses generally have a form resembling that in Figure 7.14. A very narrow, steep-sided pulse is a *spiked* pulse, while one where the corners are smoothly rounded is a rounded pulse.

The pulse dimensions shown in Figure 7.14 are based on the *peak amplitude,* which is the maximum amplitude that can be measured vertically from the baseline to any point on the pulse. You should not confuse this with the *pulse amplitude,* which is the amplitude of the *centrum* above the baseline.

The *rise* and *fall times* of the pulse indicate its steepness. The steeper the rise time, the more likely is the leading edge to overshoot and cause a ripple in the first part of the pulse top. Such an overshoot and ripple (called *ringing*) are known as *pulse top distortion.* Since many amplifiers will cause similar distortion to an otherwise good pulse, you need to know what the pulse looked like before it was applied to the amplifier, for comparison.

The pulse equivalent of frequency (number of pulses per second) is the *pulse repetition rate* (PRR); it is wrong to speak of "pulse frequency" because we are not in the frequency domain. Pulses, lacking the unbroken form of a continuous wave, have much more the nature of chopped DC, and to attempt to apply frequency-domain mathematics to them doesn't work too well. The exception is the square wave, which is a continuous wave having many pulse characteristics also.

The block diagram of a pulse generator in Figure 7.15 shows that the pulse originates in the *repetition rate oscillator,* unless an external trigger is used. The repetition rate oscillator is an astable oscillator, such as a free-running multivibrator. The rate is determined by the setting of the repetition rate control, which selects the proper fixed capacitor for the desired range. This capacitor charges through a resistor until its voltage, which is coupled to the base or emitter of a transistor, causes the latter to conduct and discharge the capacitor. A vernier control is provided by connecting a variable capacitor in parallel with the fixed one, or a variable resistor in series with it.

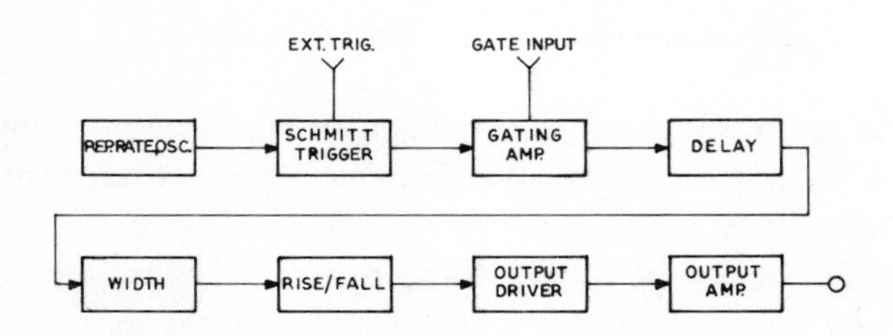

Figure 7.15 Block Diagram of Pulse Generator

The signal put out by the repetition rate oscillator is used to activate the *Schmitt trigger*, or an external source may be used. The Schmitt trigger is a single-shot multivibrator that produces an output of fixed amplitude and duration for each input trigger pulse, regardless of its shape, as long as it has the proper polarity and sufficient voltage.

Pulses may be produced continuously or in bursts. The *gating circuit,* which is a bistable multivibrator, or flipflop, is driven by the signal from the Schmitt trigger when the gate-mode switch is in the non-gated position. When this switch is rotated to the synchronized position, input synchronizing pulses are applied to the repetition rate oscillator, so that its rate is controlled externally. When in the asynchronized position, external input pulses are applied to the gating circuit, and turn it on for the duration of each. In this way, bursts of pulses are "gated" through the *gating amplifier*.

A second pulse is generated in the *delay circuit*. This circuit consists of two multivibrators connected by an RC circuit. The pulse from the gating circuit goes to the first multivibrator and switches it to its other state; its output pulse then charges the capacitor of the RC circuit. When this reaches the voltage required to switch the second flipflop, a second pulse is produced, the interval between them depending upon the RC constant. This can be adjusted by the capacitor, which is variable. One of the pulses can be eliminated by switching off its multivibrator, so you have a choice of single or double pulses. Double pulses are useful for testing memory cores and counter circuits.

The *width circuit* operates in the same way as the delay circuit, except that it is used to adjust the pulse width.

The *rise* and *fall time circuit* gives the pulse its final shaping. The slopes of the leading and trailing edges of the pulse are set by RC circuits in which the range is selected by choice of capacitor, charging through rheostats that act as verniers. The slope varies with the current, so adjusting the current will steepen or ease the slope.

The *output amplifier* is a Class A amplifier, so as to give an undistorted output. It usually consists of two channels, to give two outputs, one positive and one negative. An amplitude attenuator of the bridged-T type adjusts the amplitude of each channel.

OPERATION

The following step-by-step procedure will guide you in the operation of low-frequency signal sources. Some additional procedure is required for higher frequencies.

1. Unless battery-operated, connect to suitable power source and turn power switch on. Allow sufficient time for oscillator to stabilize. This should be about 30 minutes for vacuum-tube equipment, less for other.
2. Set range and frequency controls for desired frequency. Frequency dial setting must be multiplied by range switch setting. For instance, 45 (on frequency dial) x 10 (range setting) = 450 hertz.
3. Set function switch (if applicable) for the waveform you want.
4. Connect equipment under test to output terminals of oscillator, using matching pad or attenuator as necessary. (If a very small output voltage is required it is better to use an external attenuator to reduce the ratio of hum and noise.)
5. Adjust amplitude control for output voltage required. This may be monitored at the input of the unit under test with an AC voltmeter if accuracy is required.
6. The output terminals of low-frequency oscillators and generators consist of two or three binding posts. One of them (often colored red) is for the output from the high side of the output amplifier, the other is the ground return and connects to the chassis. If the output amplifier is a differential amplifier (see Chapter 1), there will be an additional terminal, because the amplifier has two high sides. This terminal may be connected to ground with a shorting strap between it and the ground terminal, to give an unbalanced output, or not so connected, to give a balanced output.

   These terminals are normally spaced ¾ inch apart so you can plug in a "double-banana" plug.

For higher frequencies, and especially to preserve the shape of fast-rising pulses and square waves, you have to take additional steps to make proper connections between your generator, the circuit under test and other equipment connected to them. The higher the frequency and the faster the risetime, or the longer the connecting leads, the more important this is. Shielded or coaxial test leads, with suitable matching pads, as described above under Signal Generators, are definitely required.

NOISE GENERATORS

Nature of Electrical Noise

Natural electrical noise originates in many ways. For instance, when current flows through a carbon resistor the haphazard movements of electrons resemble those of pebbles in a mountain torrent. These innumerable random impulses, when amplified, give us the phenomenon we know as noise. When made audible in a speaker or headphones, it sounds like something frying. On an oscilloscope screen, it looks like the side elevation of an unkempt lawn.

Unlike a sine-wave signal in which each cycle has the same amplitude, frequency and phase as the others, noise impulses are individually unpredictable. Their characteristics have to be determined by taking the average of large numbers of them. Consequently, we think of noise as having a bandwidth or spectrum, with its energy distributed over this spectrum. This is called its *power density spectrum* (PDS).

There is some analogy here to the visible spectrum of light, so that noise that has the same average power density all across its spectrum is called *white noise*. It has "constant energy per hertz bandwidth." However, if we pass white noise through a filter that progressively attenuates it toward the high end of the band it will be like white light passing through a light filter that is progressively more opaque toward the violet end of the spectrum. Such a filter would be a pink filter, and white light would be pink after passing through it. White noise after passing through a pink noise filter becomes *pink noise*.

Pink noise has "constant energy per octave bandwidth," which is another way of saying its energy level decreases by half (minus 3 dB) each time the frequency doubles. You can also have a filter that attenuates on each side of a standard frequency, to give an energy-level peak at that frequency. The resulting output is known as *ANSI noise* (ANSI = American National Standards Institute, Inc.). White, pink and ANSI noise are shown in graph form in Figure 7.16.

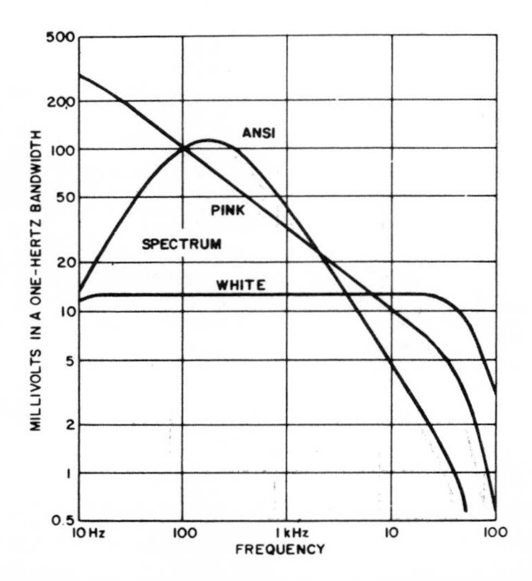

Figure 7.16 White, Pink, and ANSI Noise Spectra *(After General Radio)*

The amplitude of each individual impulse is also random. However, naturally-occurring phenomena generally are distributed so that they follow a Normal (Gaussian) curve. This curve is bell-shaped, and centered on the average value. It is called the *probability density function* (PDF) of the noise signal. It has absolutely nothing whatever to do with the power density spectrum.

Noise-Signal Generators

Most noise-signal generators are random-noise generators, and are fairly simple instruments. The noise source is a naturally ''noisy'' component such as a carbon resistor, gas-discharge tube or suitable semiconductor. Its output is amplified and filtered to give the desired noise spectrum.

Most generators have a Gaussian PDF, and supply white, pink or ANSI noise. A white-noise-only generator can be used with a pink noise filter to give pink noise also. They are used to study the effectiveness of receivers in detecting and recovering signals in noise; the frequency response of circuits (in the same way as a sweep generator); simulate speech, music or communications circuit traffic; drive random vibration equipment; and study acoustic and psychoacoustic problems (white noise has been used with some success as an anesthetic in dentistry).

8

ELECTRONIC COUNTERS

Counters use binary arithmetic (see Chapter 1). This means that the circuits of a counter can be in one or the other of two states. Transistors are either conducting or not conducting, so that specific voltages are present or absent. The binary method of notation is used to symbolize these conditions: 1 represents the presence of a voltage, 0 its absence.

The switching on or off of a counter's transistors takes place at electronic speeds. In the following description, we speak of a circuit being in the 1 state or the 0 state, as if it were a light bulb that is lit or not lit. This is for our convenience, because we've got to slow things down to a speed our minds can keep up with. However, we should always remember that the circuits of a counter such as that in Figure 8.1 are designed to switch millions of times a second. Consequently, these DC voltage states become high-speed trains of pulses, with the characteristics of pulses described in the previous chapter.

Figure 8.1 This electronic counter can count frequencies up to 50MHz. The reading will continue to be displayed after the input signal is disconnected if the FAST-NORM-HOLD switch is placed in the HOLD position, as shown. *(Courtesy Hewlett-Packard Co.)*

The heart of an electronic counter is a set of *decade counting units* (DCUs), one for each figure displayed in the readout. Each DCU is made up of flip-flops (bistable multivibrators) cascaded in such a way that when one changes all the others will be affected. Each flip-flop is a *binary counter* that is worth 1 or 0 according to its state. If we start out with four flip-flops all in the 0 state and apply a triggering pulse to the first flip-flop, it will change to the 1 state. A second pulse will return it to the 0 state, but in so doing it triggers the second flip-flop to the 1 state, and so on with succeeding pulses.

Binary Counter

Figure 8.2 shows the count sequence of a binary counter constructed with four flip-flops. As each pulse is received, the flip-flops change state accordingly. At the fifteenth they are all in the 1 state. The sixteenth pulse resets them all to zero.

PULSE NUMBER	STATE OF FLIP-FLOPS				DECIMAL EQUIVALENT OF EACH BIT				DECIMAL TOTAL
	A	B	C	D					
0	0	0	0	0	0	0	0	0	0
1	1	0	0	0	1	0	0	0	1
2	0	1	0	0	0	2	0	0	2
3	1	1	0	0	1	2	0	0	3
4	0	0	1	0	0	0	4	0	4
5	1	0	1	0	1	0	4	0	5
6	0	1	1	0	0	2	4	0	6
7	1	1	1	0	1	2	4	0	7
8	0	0	0	1	0	0	0	8	8
9	1	0	0	1	1	0	0	8	9
10	0	1	0	1	0	2	0	8	10
11	1	1	0	1	1	2	0	8	11
12	0	0	1	1	0	0	4	8	12
13	1	0	1	1	1	0	4	8	13
14	0	1	1	1	0	2	4	8	14
15	1	1	1	1	1	2	4	8	15
16	0	0	0	0	0	0	0	0	0

Figure 8.2 Count Sequence for Four-Bit Binary Counter (which counts to 15, then returns to 0).

Each flip-flop in the 1 state is storing a binary number. Its decimal equivalent is 1 if in A, 2 if in B, 4 if in C and 8 if in D. For instance, when the thirteenth pulse has been received the state of the flip-flops is as follows:

Flip-flop:	A	B	C	D
Binary number:	1	0	1	1
Decimal equivalent:	1		4	8, which add up to a total of 13.

BCD Counter

Since we all use the decimal system, it is more convenient to have our counters read out in that system. For this we need DCUs that count from 0 through 9 and then recycle. We can use the same flip-flops as in the binary counter but they must all reset to zero on the tenth pulse. Figure 8.3A shows the diagram of a *binary-coded decimal* that does this with four JK flip-flops and two NAND gates.

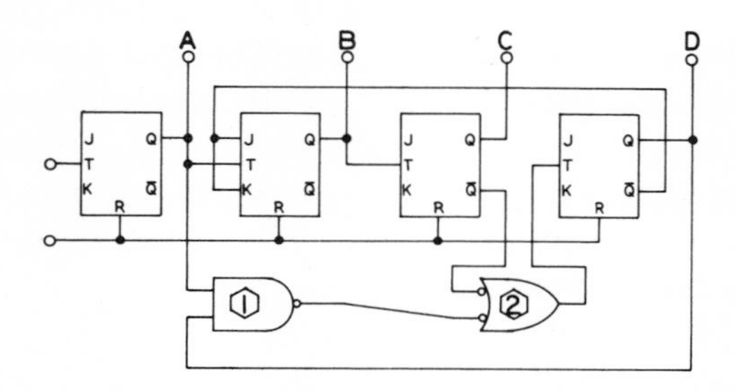

Figure 8.3a BCD counter using four JK flip-flops and two NAND gates. The outputs labeled A, B, C, D are connected to the readout decoders. The R inputs are connected to a RESET push-button (see Figure 8.1) which enables you to reset all the flip-flops to zero at any time.

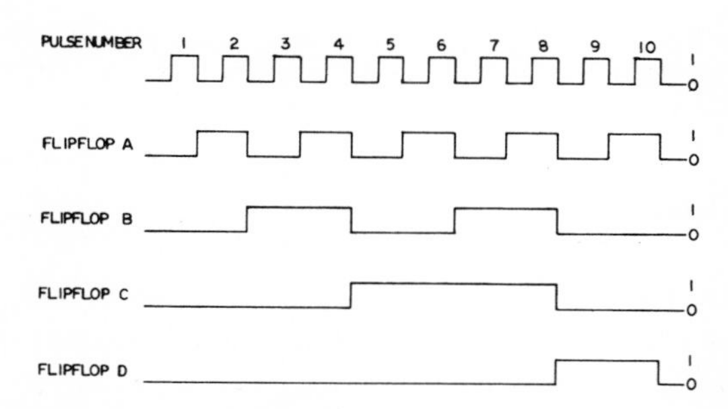

Figure 8.3b Each flip-flop changes state, from 1 to 0 or from 0 to 1, on arrival of the *trailing* edge of the input pulse.

JK Flip-flops

A JK flip-flop, as shown in Figure 8.4, consists of eight NAND gates and one NOT gate (inverter). C and D, G and H, are connected to form two reset/set (RS) flip-flops. They are clock-synchronized by the addition of the NAND gates A, B, E and F. One of the inputs of each of these is tied to the common clock or trigger line from input T. This means they cannot change state state until a positive clock or trigger pulse is applied.

As you can see, there are two halves to this IC. That to the left (A, B, C, D) is the *master* flip-flop, that one to the right is the *slave*. Whatever is at the J and K inputs is transferred to the master flip-flop by the leading edge of the trigger pulse at T. At the end of the trigger pulse (on its trailing edge) the information in the master is passed to the slave flip-flop, and is reflected in the output Q and $\overline{Q}$. (The bar over the Q means that the $\overline{Q}$ output is the reverse of the Q output. Pronounce it "not Q," "negated Q," "inverted Q," or "Q's complement," whichever you prefer.)

The effect of all this is that if the input signals at J and K are both 1, or are not the same as those at Q and $\overline{Q}$, the output signals will reverse when a trigger pulse is applied at T.

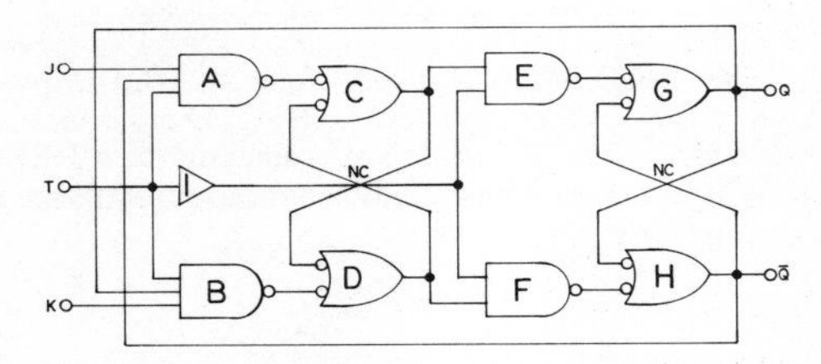

Figure 8.4 A JK flip-flop consists of a "master" and "slave." Supply voltage and reset connections are not shown. (NC = no connection.)

BCD Counter Operation

Let's see how the BCD counter in Figure 8.3A works. The J and K inputs of flip-flops A, C and D do not have signal connections. They are connected to a positive voltage source, so are in a 1 condition at all times. These flip-flops will switch on the trigger pulse alone. The J and K inputs of B are connected to the $\overline{Q}$ output of D. As long as this is 1 the B flip-flop will also be "enabled."

When a pulse arrives at the A flip-flop's T input, the A flip-flop's Q output changes from 0 to 1 on the trailing edge of the pulse, as explained above. This output does not trigger the B flip-flop until the next pulse into the A flip-flop switches its Q output back to O. B then switches, and the input to C becomes a 1. Notice that it took *two* input pulses to get *one* output pulse from A. It takes two input pulses into B (from A) to get one out of B, and so on, as you can see by looking at Figure 8.3B. This IC can also act as a frequency divider.

B's output to C causes its Q output to switch also, but this is not connected to D. Let's see why. We've had seven pulses applied to A's input, and flip-flops A, B and C are all in the 1 state. Now we want to get these back to O, and D from the 0 state to a 1.

The $\bar{Q}$ output of C is 0, and is connected to one input of NAND gate 2. These NAND gates, though drawn differently, both work the same way. The output will be a 1 for all inputs except a 1 on both. Since the output of NAND gate 1 is a 1, the output of NAND gate 2 is a 1 as long as the $\bar{Q}$ output of C is 0. When the either pulse arrives at A, A goes 0, turning B and C off in turn. C's $\bar{Q}$ output is now 1, so NAND gate 2 now has two 1 inputs, so its output goes 0. This change of D's input from 1 to 0 causes it to switch, and its Q output becomes 1. The condition of flip-flops A, B and C and D is now 0 0 0 1 (= 8). D's Q output is also applied to NAND gate 1, so when A next goes 1 it will switch and reset NAND gate 2 to its former condition of a 1 output.

On the arrival of the ninth pulse, A's output switches to a 1, so both inputs of NAND gate 1 are 1, and its output is 0. This resets NAND gate 2 to a 1 output.

When the tenth pulse arrives, the A flip-flop goes to zero. This cannot affect the B flip-flop because the J and K input are both 0. A's Q output changes one of NAND gate 1's inputs to 0, so its output changes to 1. This puts a 1 on both inputs of NAND gate 2, so its output becomes a 0. This causes D to switch to 0, so all four flip-flops have now been reset to zero.

Incidentally, each flip-flop has an R input also, and all four R inputs are connected to a common line. When a signal is applied to this line, all the flip-flops are reset to zero. Most counters, including those in Figures 8.1 and 8.7, have a RESET push button on the front panel, so you can clear the counter and return all its DCUs to zero whenever you want to.

Since each BCD counter counts from zero through nine, you can use them to count any size of number by "daisy-chaining" them, with one DCU for each digit in the readout, as shown in Figure 8.5. As the tenth pulse causes flip-flop D in the units DCU to return to the 0 state, it causes flip-flop A in the tens DCU to change state to 1. In this way the tens DCU counts one for every ten counts of the units DCU. A 1 is carried forward

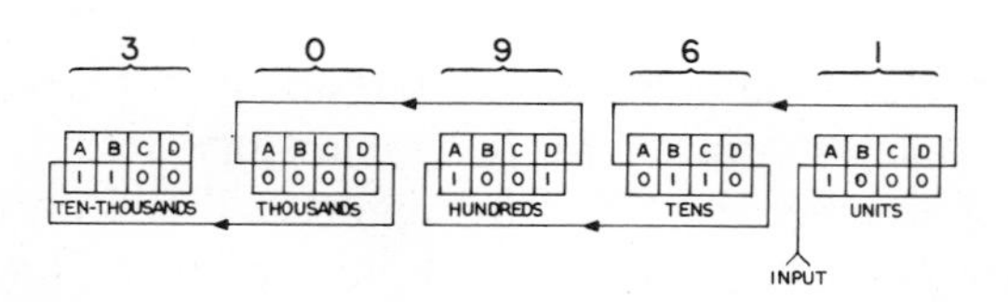

Figure 8.5 How five DCU's are used to give a five-figure read-out.

to the next DCU to the left each time the DCU to its right counts to ten, just as in simple addition.

In Figure 8.5, the state of each DCU is shown in its binary form, with its decimal equivalent above. The latter is what we want to appear in the readout.

There are various kinds of readouts. The counters in Figures 8.1 and 8.7 use a row of "nixie" lamps. These are gas glow tubes with wire figures 0 through 9 inside. A *decoder* energizes the proper connections to make the figure glow that corresponds to the number stored in the DCU. Other types of display that operate similarly are electroluminescent, light-emitting diode (LED), liquid crystal and cathode-ray tube (CRT).

Complete Counter

A complete counter, such as the one in Figure 8.1, typically contains the circuits shown in Figure 8.6. These are:

Input amplifier	DCUs
Schmitt trigger	Internal time base oscillator
Main gate	DDAs

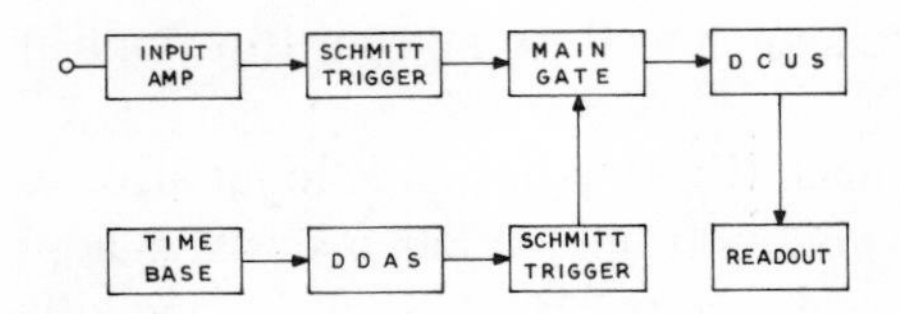

Figure 8.6 The Circuits of a Complete Counter.

Input Amplifier and Schmitt Trigger

The counter must be able to count whatever you want it to, be it sine waves, pulses or even random impulses, within its frequency range. However, the DCUs can only operate reliably on trigger pulses of suitable dimensions. Therefore, the signal at the input is first amplified (an input attenuator determines the amplification factor) and then shaped by the Schmitt trigger. (In some counters the circuits may be combined.) The amplifier also contains controls for level and slope selection. With them you can control whether the Schmitt trigger shall fire on the leading or trailing edge (negative or positive slope), and at what voltage level of the input signal. Similar circuits are used in oscilloscopes, discussed in the next chapter (9).

Main Gate

The main gate opens for a period of time determined by the setting of the *time-base control*. How often it opens is set by the *sample rate control*. The number of pulses counted by the DCUs depends upon the frequency of the input signal and the length of the time for which the gate is open. If this is set to one second (with the function control set to *frequency*), then the number counted will be "per second," which is the frequency in hertz, of course.

However, when the function control is set to *period,* the input signal opens and closes the gate, and the DCUs count the time between pulses by counting the number of *oscillator* pulses received while the gate is open. This is the more accurate way to measure the frequency of very low frequency signals ($f = \frac{1}{t}$).

Oscillator

The time base may be internal or external: from an internal oscillator or an external time standard. An internal oscillator is used in most cases, and is generally a crystal oscillator. Greater accuracy is obtained by mounting the crystal in an oven controlled by a thermostat. It goes without saying that after turning on the power an adequate time must be allowed for the crystal to reach its operating temperature and stabilize.

DDAs

The oscillator signal has a fairly high frequency (100 kilohertz or 1 megahertz) which is then subdivided to give the period for which the main gate will be open. This is done by the *decade divider assembles* (DDAs). Each DDA divides by ten, as in a DCU (which, you'll recall, we said could be used as a frequency divider), so that in some counters they are called DCUs anyway. The time-base control selects how many DDAs will be placed in series with the oscillator, so determining whether the oscillator signal will be divided by 10, 100, or 1000, or not divided at all.

Some counters have inputs for two signals. When the function switch of one of these counters is set to "ratio" and different signals are applied to each input, one signal controls the main gate and the other is counted by the DCUs. The readout then gives the ratio of the one to the other, multiplied according to the setting of the time-base control.

MEASUREMENT ACCURACY

The accuracy of most counters is given in the form "±1 count

±time-base accuracy ±trigger error." This is a little different from the way in which the accuracies of voltmeters are expressed, so let's see what it means.

±1 count refers to the ambiguity found in any conventional counter, and is not important except at very low frequencies. It means that because the input signal and time base are not synchronized a chance variation of one count may occur in consecutive samples. This is why it is better to count the period between events with very low frequencies, because the number of events occurring in the unit of time is small. The more increments you count, the less important is an error of one.

Time-base accuracy is simply the accuracy of the oscillator. Where the counter uses an external time base, such as a cesium-beam standard, it can be very high (7 parts in 10^{12}). However, most of the counters you will encounter will be using an internal quartz-crystal oscillator.

Time-base accuracy is divided into:

Long-term stability   Temperature and line-voltage variation
Short-term stability

Long-term stability is the rate at which the crystal ages, with consequent gradual change of frequency. It is expressed by a statement such as: "3 parts in 10^7 per day at 25°C." This would mean that at the end of a month the cumulative drift of the oscillator frequency due to crystal aging would be 30 (3 parts in 10^7) = 9 parts in 10^6, or 9 parts per million. A one-megahertz signal would be off by nine hertz. This error is removed by periodically recalibrating the oscillator. The interval between calibrations determines the accuracy of the counter, as far as long-term stability is concerned.

Short-term stability is the degree of instability and noise going on in the oscillator all the time, which could affect the accuracy of the time base. The effect will be greater for short gate times. It is often given for an average of one second, expressed in the same way as long-term stability. Another name for short-term stability is *fractional frequency deviation*.

Temperature and line-voltage variation effects are similar to those met with in other electronic equipment. They are minimized by using crystal ovens, controlled ambient temperature and regulated power supplies.

Trigger error occurs only in period measurements, when the input signal controls the main gate. Noise on the signal may cause the gate to open or close incorrectly. It is worst for noisy, low-frequency signals.

TYPES OF COUNTERS

The counter illustrated in Figure 8.1 is a "universal" or general-purpose counter. We also mentioned counters with two inputs that could show the ratio between two frequencies. Two other types are the *preset counter* and the *plug-in counter*.

Preset counters have a set of decade thumbwheels on which you can set a total number of events to be counted. The counter accumulates the input signals until this number is reached, when it is enabled to send a control instruction or signal to other equipment, to control some process. Preset counters are used for batching, and also for the precise control of weight, liquid level, length, and other parameters in industrial processing.

Plug-in counters are general-purpose counters whose versatility is increased by the use of plug-in units. The basic counter is similar to the universal counter in Figure 8.1, and is called the *mainframe*. However, a portion of its front panel is removable. When you detach this, you uncover a receptacle in which you can install any one of a range of *plug-ins*. On the rear of each plug-in is a male connector that mates with a female connector at the back of the receptacle when the plug-in is fully inserted. Its front panel is then flush with the front panel of the mainframe. Each plug-in panel has its own controls according to its function. Principal types of plug-in are:

Heterodyne converter	Time Interval unit
Transfer oscillator	Preset unit
Prescaler	Voltage-to-Frequency converter

A *heterodyne converter* plug-in is a mixer that operates in the same way as the "front end" of a TV or radio set. It mixes a local oscillator signal with the frequency to be measured to give an IF signal within the frequency range of the mainframe. In this way, a basic counter that can count up to 50 megahertz (such as that in Figure 8.1), if constructed to take plug-ins, could have its range extended to as high as 18 gigahertz. The plug-in takes the counter's own time base and multiplies it. The resulting higher-frequency signal is then applied to a harmonic generator, and the appropriate harmonic is selected to be mixed with the signal to be counted. The difference frequency then goes into the mainframe to be counted and displayed on the readout. A frequency control on the front of the plug-in enables you to select the mixing frequency. The dial setting of this control added to the frequency displayed in the readout gives you the unknown frequency.

A *transfer oscillator* plug-in works in a similar manner, except that the plug-in contains a local oscillator that is tuned for a zero beat with the unknown frequency. This must, therefore, be a harmonic of the unknown frequency. The local oscillator frequency is counted by the counter and multiplied by the ratio the unknown frequency bears to it (depending upon which harmonic is used).

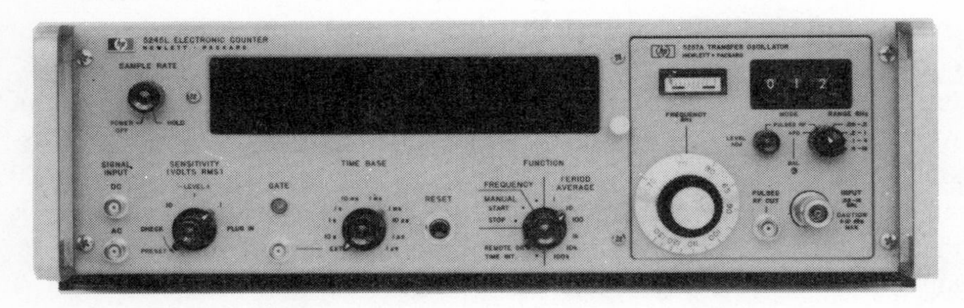

Figure 8.7 A plug-in counter with a transfer oscillator plug-in that extends its frequency range to 18 GHz. The SENSITIVITY control (bottom, left-hand corner) in addition to being an input attenuator, also has a position to select the plug-in. The small red knob on this control is independent of it, and is the LEVEL control. When turned to the PRESET position, the counter triggers at the level that is optimum for signals that are symmetrical about ground.

In the example shown in Figure 8.7, you can see the local oscillator tuning dial, the zero beat (''phase lock'') meter and a set of thumbwheel switches. If you set these to the number of times the ''unknown'' frequency is greater than the local oscillator frequency, the counter will give a direct readout of the ''unknown.''

A *prescaler* plug-in also extends the frequency range of the counter, but not as much as the heterodyne converter and transfer oscillator plug-ins. It works by dividing the input signal down to a frequency within the counter's range, and increasing the main gate opening time by the same factor, so that the displayed frequency is that of the signal being measured.

A *time-interval unit* plug-in has two separate input channels, START and STOP. When an external signal is applied to the START channel, the main gate is opened, and the DCUs count clock pulses from the internal oscillator. This continues until an external signal is applied to the STOP channel, when the main gate closes, and the display shows the total number of pulses.

Each channel has *trigger level* and *slope* controls that can be set so that you can measure the time interval between any two points on the input signal. For example, if you want to measure the rise time of a pulse (as defined in the Chapter 7), you connect the signal to both inputs. On the START controls you set *positive* slope, and trigger level at 10 percent of the pulse amplitude voltage. On the STOP controls you set *positive* slope, and trigger level at 90 percent of the pulse amplitude voltage. Each time the voltage reaches 10 percent on the pulse leading edge, the main gate opens; it closes when the voltage reaches 90 percent. The number of clock pulses counted and displayed indicates the interval. If the *time-base* control was set at 1 micro-second, and the readout is 22, it means that 22 time-base pulses were counted between START and STOP, so that the risetime must be 22 microseconds.

Preset plug-ins convert universal counters to preset counters as described above.

Voltage-to-frequency converter plug-ins change a voltage to a frequency as in an integrating digital voltmeter (see Chapter 6) so that the counter becomes a DMM for the time being, reading voltage instead of frequency or time interval.

9

HOW TO USE THE OSCILLOSCOPE

If electronic instruments were chessmen, the oscilloscope would be the Queen, for it can duplicate all the measurements of voltage, current, resistance, frequency, phase, and so on, performed by other instruments; and in addition it can enable you to evaluate rapidly-changing phenomena, including those events that happen only once and last only nanoseconds.

The peculiar power of the oscilloscope lies in its ability to produce an electronic picture to show the variation of one quantity with regard to another. These quantities are electrical signals, as found in a circuit or obtained from a transducer. The use of transducers makes it possible for the oscilloscope to measure pulse rate, analyze engine performance, or portray many other non-electrical quantities. However, oscilloscopes mostly are used to display variations in voltage amplitude over a period of time.

In this respect, the oscilloscope is performing a function similar to a barograph or seismograph, in which a pen traces an ink line on a chart that moves forward at a steady rate. In the case of electrical signals, however, they are mostly repetitive: the same amplitude variation is repeated continuously at a fixed rate. For instance, a train of pulses may have a repetition rate of one kilohertz. The period of time from a point on the leading edge of one pulse to the same point on the next is, therefore, one millisecond. If we set our oscilloscope's "pen" to travel across its "chart" in one millisecond, while at the same time it rises and falls in accordance with the amplitude variations of the pulse, it will draw a picture of the pulse, as in Figure 9.1.

Because it lasts only a thousandth of a second we can't see this picture, unless we "freeze" it, as in a photograph. However, if we make the "pen" repeat its sweep across the "chart" every millisecond, in synchronization with the pulse train, successive pulses will be superimposed continuously to form a visible image. It will seem as if the signal is standing still on the screen, as a spinning wheel seems stationary under a synchronous strobe light.

A mechanical pen is limited by its mass to a maximum frequency response of around 150 hertz. But the electron-beam "pen" in the oscilloscope cathode-ray tube is practically weightless, and is capable of responding to frequencies beyond a gigahertz.

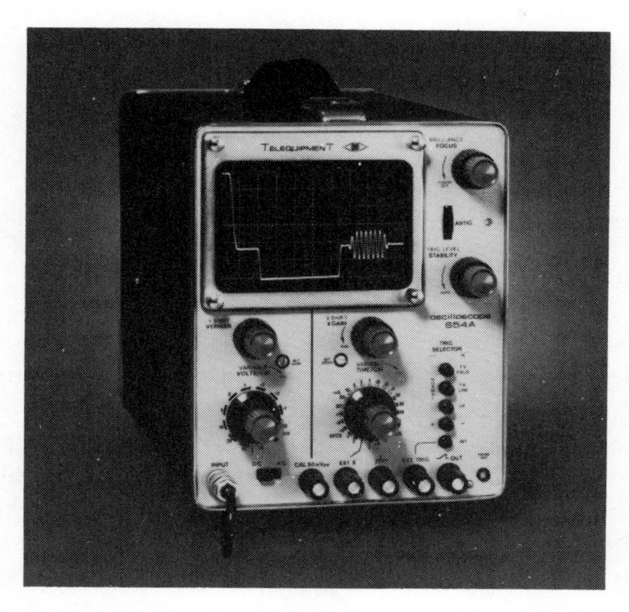

Figure 9.1 This is a moderately-priced oscilloscope, with a bandwidth from DC to 10MHz, and maximum vertical sensitivity of 10mV per cm. It is designed for general use, but is slanted toward TV, as is seen by the special TV triggering inputs that may be selected. It is displaying a horizontal sync pulse and color burst. *(Courtesy Tektronix, Inc.)*

CATHODE-RAY TUBE

The cathode-ray tube (Figure 9.2) is the heart of the oscilloscope. This is a funnel-shaped vacuum tube made of glass, or ceramic with a glass screen. The narrow end is the base as in a TV picture tube, which it resembles. This end houses an electron gun consisting of heater filament, cathode, grid, and focussing and accelerating anodes, which direct a narrow, high-velocity electron beam at the center of the screen at the other end.

The screen is a thin layer of fluorescent material, called phosphor, which coats the inside surface of the glass. In most modern oscilloscopes, this part of the envelope is rectangular and flat, with a *graticule* etched in the phosphor to provide a scale for measurement of the quantities displayed.

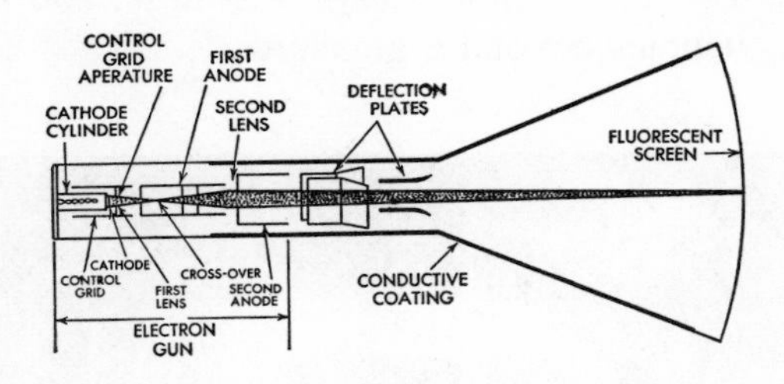

Figure 9.2 Cathode-Ray Tube. The CRT is like a black-and-white picture tube, except that it uses deflection plates instead of yoke coils so that it can handle a wide range of sweep rates.

Several different phosphors are in service. The one most commonly used gives a yellowish-green fluorescence wherever the electron beam strikes it with sufficient force.

After leaving the accelerating anode, the electron beam passes between two pairs of *deflection plates*. The first pair is for vertical deflection. A positive potential on the upper plate, with a corresponding negative potential on the lower one, bends the beam upward, so that it strikes the upper part of the screen. The amount of deflection is according to the potentials on the plates, which are driven by the push-pull output of the vertical amplifier. The second pair of plates is driven in the same way by the horizontal amplifier, except that the deflection is from side to side.

The electron gun described above is called the *writing gun,* because it produces the electron beam that traces out displayed waveforms in obedience to the deflection-plate potentials. The electron beam is accelerated by anode potentials as high as 24 kilovolts in some oscilloscopes, so it packs a lot of energy. Where it strikes the screen it gives up this energy, making it glow and dislodging a great many secondary electrons from the phosphor. The "written" area from which the electrons were dislodged acquires a positive charge. The dislodged electrons are collected on a conductive coating on the inside of the tube, from which they travel to the positive side of the high-voltage supply.

In standard CRTs, this written area soon replaces its missing electrons and becomes neutral again. In *storage tubes,* however, it can be "frozen" so that the trace remains on the screen for up to an hour. This is done by the addition of *flood guns,* as shown in Figure 9.3. The flood guns emit low-velocity electrons that strike the unwritten areas of the screen with too little energy to cause them to glow or emit secondary electrons. The flood-gun electrons arriving first at the screen build up a negative charge on the phosphor that repels following electrons. But when flood-gun electrons approach a *written* area, its positive charge attracts and accelerates them so that they impact with sufficient energy to make the trace glow and maintain its positive charge by dislodging more secondaries.

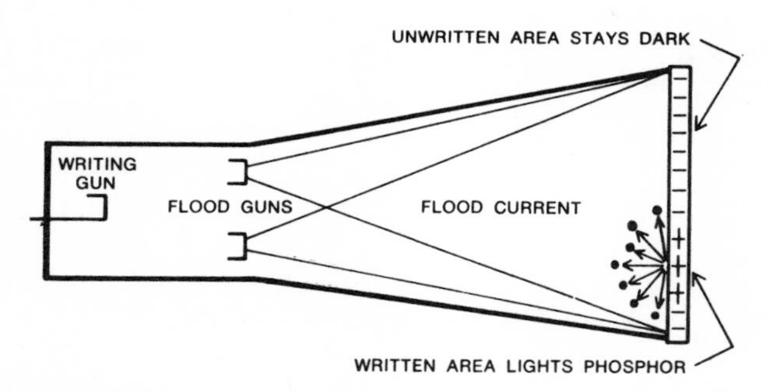

Figure 9.3 Flood gun electrons hit unwritten areas too slowly to light phosphor and the target charges negative. The positively charged written area attracts electrons at high speed keeping the phosphor lit and dislodging enough secondaries to hold the area positive. *(Reprinted by permission of Tektronix, Inc.)*

The storage tube enables you to store ''single-shot'' phenomena for study or photography. For these one-time events a control is provided to *enhance* the energy of the writing gun electrons, so that a bright recorded display is produced with only one sweep.

The flood guns are usually arranged so that you can use all or half of the screen for storage. The split-screen display allows you to compare a second signal with a previously-stored one. The writing gun may also be used to overwrite already existing displays in the storage area. This is called ''write through.''

When you switch the flood guns off, a storage tube becomes a conventional CRT with no storage.

SUPPORTING CIRCUITS

To enable the CRT to function, modern oscilloscopes use the circuits shown in block form in Figure 9.4. The delay line has a dashed outline because it is usually omitted in instruments that handle only low frequencies.

You feed the waveform you wish to observe to the vertical amplifier. This has a *sensitivity control* that works in the same way as the range switch on a VTVM (see Chapter 6). It reduces the amplitude of the input signal, so that after amplification it has a size that will fit on the CRT screen. The dial of this control is marked in volts and millivolts per vertical scale division. This means that if you set it at 1 volt per division, for instance, a one-volt input will cause the electron beam to be deflected one graticule division up or down, depending on whether the voltage is positive or negative.

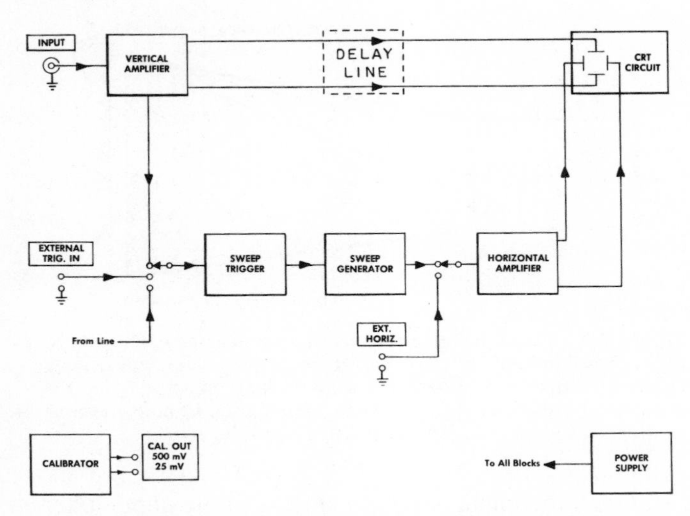

Figure 9.4 Oscilloscope Block Diagram. The delay line is included in some oscilloscopes to prevent the vertical signal from reaching the CRT before the sweep starts. *(After Tektronix, Inc.)*

The vertical amplifier has a push-pull output, so that one vertical deflection plate pulls, while the other pushes, the electron beam as it passes between them. In the absence of an input signal—with the horizontal deflection plates disconnected—the electron beam strikes the screen dead center, producing a stationary glowing spot. If a very low-frequency signal is now applied to the vertical amplifier input, the resultant alternating potentials on the deflection plates will cause the spot to move up and down. At higher frequencies, the spot moves too fast for your eye to follow, so it blurs into a vertical straight line.

Exactly the same thing would happen if we reconnected the horizontal deflection plates, disconnected the vertical, and applied the input signal to the horizontal amplifier; except that the electron beam would move from side to side instead of up and down.

These two directions in which the electron beam can move are also known as the Y (vertical) and X (horizontal), corresponding to the y-axis (ordinate) and x-axis (abscissa) of a rectangular-coordinate graph. Where an external input signal is applied to each amplifier (connected to its proper deflection plates, of course) the resultant display on the CRT screen is called an X-Y display, and oscilloscopes with identical vertical and horizontal amplifiers specially designed for this are called X-Y oscilloscopes. We'll come back to them later.

However, as said before, we usually want to view a waveform as a graph of amplitude variations plotted against time. Consequently, we apply an internally-generated signal to the horizontal amplifier to sweep the electron beam across the screen from left to right at a known rate. While it is doing this, it will also respond to the fluctuating potentials on the vertical deflection plates caused by the waveform's amplitude variations.

This sweep is formed by applying voltages to the horizontal deflection plates, so that at first the left-hand plate is at a positive potential and the right-hand plate is at an equally negative potential, bending the electron beam to the left-hand margin of the graticule. Then the voltage on the left-hand plate is steadily reduced until it is eventually as much negative as it was positive before. At the same time, the right-hand plate goes from negative to positive. This drives the electron beam at an even rate across the screen until it reaches the right-hand margin of the graticule. At this point, the deflection-plate potentials switch back to what they were at the start, and the electron beam returns to the left-hand edge of the screen, ready for the next sweep.

The equal and opposite voltage ramps on the horizontal-deflection plates are produced by phase-splitting and amplifying the single-ended sawtooth output of the time-base generator. This sawtooth signal is produced by charging a capacitor through a resistor, as in other RC circuits, using precision components. By choosing the proper component values any charging rate can be obtained, so a selection of resistors and capacitors is mounted on the *sweep timing switch* to give us a choice of sweep speeds. The dial of this switch is marked in seconds, milliseconds and microseconds per horizontal scale division. This means that if you select a sweep of, say, 1 millisecond per division, the electron beam will travel across the screen at this rate. Most graticules have ten divisions horizontally, so this sweep would take ten milliseconds to complete.

If the signal you are studying has a frequency or repetition rate of 100 hertz, which is the reciprocal of 10 milliseconds, the sweep will have the same duration as one cycle of the signal. If successive sweeps can be synchronized with the signal, all the displayed waveforms will exactly coincide, giving a stable picture on the screen. An unsynchronized display will be difficult to stabilize, and will tend to drift sideways all the time.

Various circuits have been devised to synchronize the sweep with the signal. The best one, and the one used in most modern oscilloscopes, is the *time-base trigger*. This converts a portion of the input signal to trigger pulses, which are used to start each sweep.

The grid potential of the CRT is maintained at a value that will prevent electron emission except during the sweep. This blanks out retrace lines. The time-base generator turns the electron beam on at the start of each sweep, and turns if off at the end, by means of a rectangular unblanking pulse of the same length as the sweep, as for a TV picture tube.

The delay line is a ladderlike assemblage of coils and capacitors that slows the vertical signal by about a fourth of a microsecond. This gives the sweep circuits time to unblank the CRT and start the electron beam moving across the screen before the vertical signal reaches it. In this way, you see the entire waveform even though some point on its leading edge was used to trigger the sweep.

SWEEP TRIGGERING AND SYNCHRONIZING

Before the triggered sweep was introduced, the thyratron relaxation oscillator was the standard sweep generator used in oscilloscopes (Figure 9.5). The timing capacitor C charges through R until its potential reaches the firing point of the thyratron. The latter then conducts, discharging C, which starts to charge again as the thyratron is reset. The output waveform is shown at the right. You'll notice it is not very linear. Selection of values of R and C determines the sweep rate. A portion of the input waveform is applied to the grid of the thyratron to trigger the sweep. It does this by giving a sudden boost to the grid voltage just as the thyratron is approaching firing point.

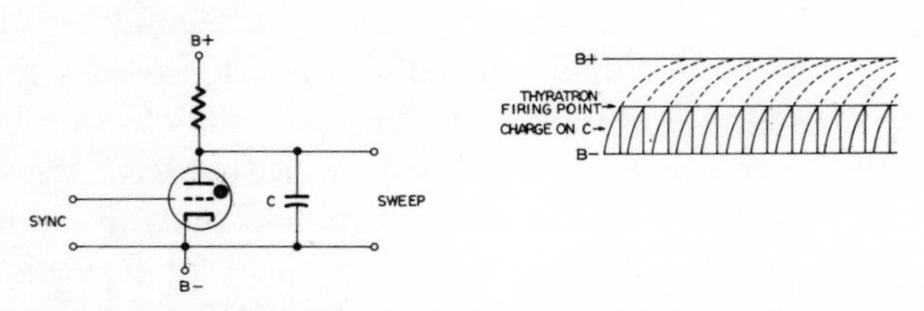

Figure 9.5 Thyatron Sweep Circuit with Output Waveform

This method has some disadvantages. It is not really linear, therefore it cannot be used to give an accurate time base. It is limited to displaying continuous recurrent signals that do not vary in frequency or amplitude. In order to match its frequency with that of the input waveform, it has to be continuously variable, so it is uncalibrated, and cannot by itself be used for measuring pulse width or frequency.

The triggered-sweep circuit consists of the subcircuits shown in Figure 9.6. The *source switch* enables you to pick off one of the following: part of the input signal from the vertical amplifier (before the delay line); an external signal via a connector on the front panel; or any other triggering signal provided for (this varies with different models). The *slope switch* allows you to reverse the input connections to the *comparator*, so as to trigger off the positive or negative slope of the waveform.

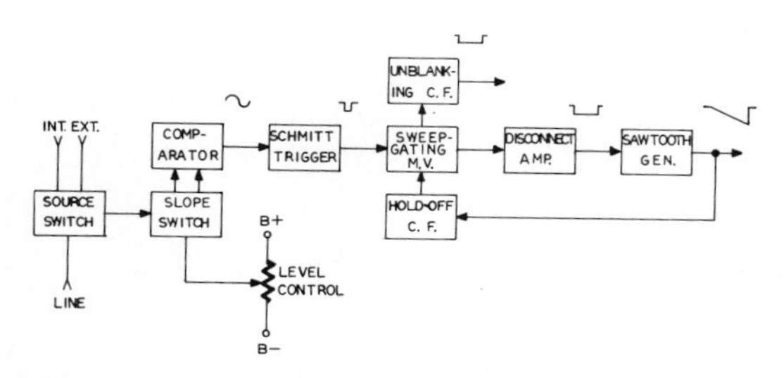

Figure 9.6 Block Diagram of Modern Triggered-Sweep Circuits

A typical comparator is shown in Figure 9.7. When you adjust the LEVEL control, you alter the conductance of V2, which changes the current through R3. This affects the cathode potential of V1, so the end result is a variation in V1's grid bias. The triggering signal is applied to V1's grid, while the output from the plate is connected to the Schmitt trigger. The value of the input voltage required to give the proper output to cause the Schmitt trigger to function will depend on the bias. Therefore, by using the LEVEL control you determine at what point on the rising (leading) edge of the waveform this will be. In other words, you are triggering on the *positive* slope at the selected voltage level.

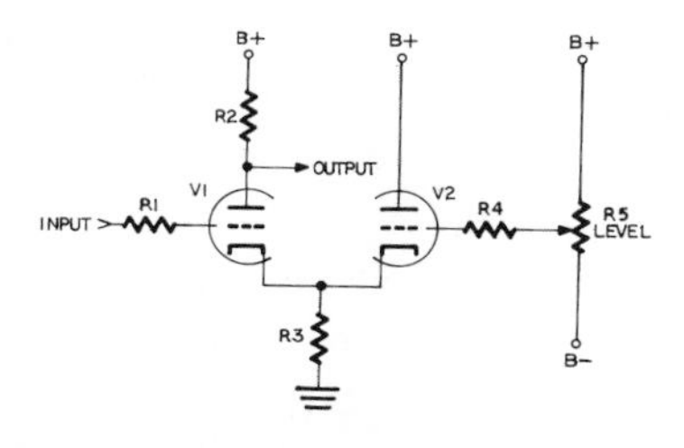

Figure 9.7 The comparator circuit selects the signal voltage level at which triggering will occur.

The slope switch reverses the functions of V1 and V2 by applying the LEVEL voltage to V1 and the input to V2. That is to say, the fixed voltage reference is now on V1's grid, the waveform on V2's. A falling voltage on V2's grid affects the balance between the two tubes in the same way as a rising one does on V1's. Therefore, the circuit is now functioning on the falling (trailing) edge. In other words, you are triggering on the *negative* slope at the selected voltage level.

The typical Schmitt trigger (Figure 9.8) consists of two emitter-coupled transistors. Q2 is normally conducting, Q1 is not. When the amplitude of the output from the comparator reaches the level that overcomes the reverse bias on Q1's base-emitter junction, Q1 switches on. This causes its collector voltage to fall, and a negative-going pulse is coupled via C1 to Q2's base, turning Q2 off. The voltage on Q2's collector rises sharply toward the supply-voltage level. When the voltage on Q1's base falls (on the trailing edge of the waveform), Q1 returns to its non-conducting state, and Q2 receives a positive-going pulse via C1 that turns it on again. The voltage on its collector drops sharply towards its former level. As you can see, this circuit will convert an input waveform of any shape to a rectangular pulse of fixed amplitude, provided the input signal's voltage rises above the level that allows Q1 to change state.

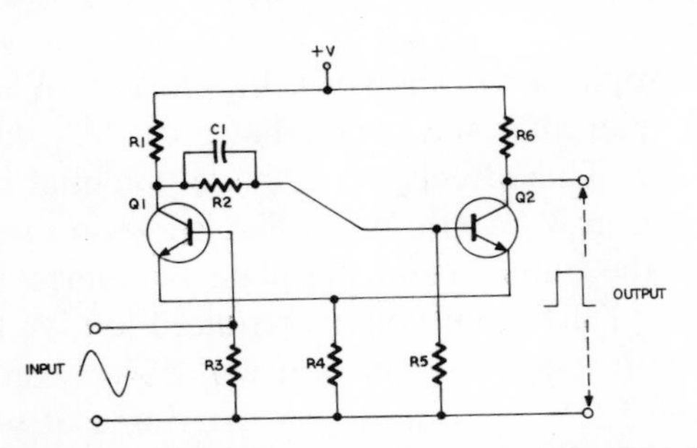

Figure 9.8 The Schmitt trigger converts an input waveform of any shape to a rectangular pulse of fixed amplitude.

The output trigger pulse goes to a *sweep-gating multivibrator,* usually another Schmitt trigger. This Schmitt trigger acts as a switch to turn off the *disconnect amplifier* (or disconnect diodes). This amplifier turns on the *sawtooth generator,* as shown in Figure 9.9.

In this example, the pulse from the sweep-gating multivibrator arrives on the base of Q1 (the disconnect amplifier), which is normally conducting. The current flows through R4, Q1, R3 and R2 to the positive side of the supply voltage (+V), maintaining a constant voltage on V1's grid. On arrival of the pulse Q1 switches off, and as C1 charges, its collector voltage starts to rise toward the value of the supply voltage. The rate of charge depends on the RC time-constant of R2, R3 and C1.

As this charge increases, the potential on V1's grid becomes steadily more positive, resulting in a rising current through R6, V1 and R7. The voltage across R7 rises also, affecting the bias on the base-emitter junction of Q2. This transistor is, therefore, increasing its conductance, so that its collector voltage is decreasing. The decrease is coupled through C1 to the grid of V1, opposing the rising potential resulting from the charge on C1. This feedback is not great enough to overcome the charge on C1, but does force it to rise at a linear rate, so that the electron beam is swept at a constant speed across the CRT screen. The sweep signal is taken in this example from the collector of Q2, and is therefore a falling voltage. This makes no difference since the horizontal amplifier to which it goes is a push-pull amplifier, as we shall see in the next section.

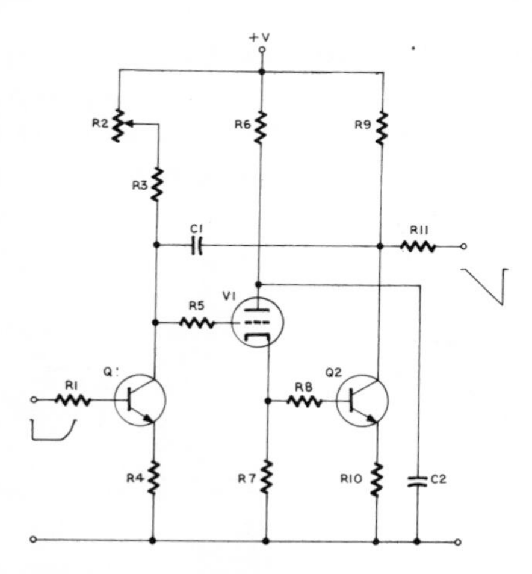

Figure 9.9 The Miller run-up circuit provides a linear sawtooth waveform.

The sweep-generator circuit is called a *Miller run-up circuit.* It can be all tube, all transistor, or hybrid, as in the example illustrated. The rate at which the sawtooth rises or falls is determined by the choice of R3 and C1. These are selectable from the precision resistors and capacitors mounted on the timing switch. R2 is a variable control used in the same way as the variable control on the vertical input attenuator. It should be kept in the fully-clockwise position (where it is switched off) if you want a fixed calibrated sweep rate.

A portion of the sawtooth output signal is fed back to the *hold-off cathode follower*. This circuit also has a resistor-capacitor pair, mounted on the timing switch. The charge on this capacitor holds the sweep-gating multivibrator from triggering on another pulse until after completion of the sweep. Without the hold-off cathode follower you wouldn't be able to show more than one cycle of a signal at a time.

When the sweep-gating multivibrator switches the disconnect amplifier off, it also switches the *unblanking cathode follower* on, switching it off again at the end of the sweep. During its "on" period the unblanking cathode follower supplies a positive voltage to the CRT grid to turn on the electron beam. Because this voltage is only applied during the actual sweep, you don't see the retrace movement of the beam as it returns to the left-hand side of the screen.

OSCILLOSCOPE AMPLIFIERS

Horizontal Amplifier

The horizontal amplifier converts the single-ended sawtooth output from the sawtooth generator into a push-pull signal suitable for driving the horizontal deflection plates. In this it is similar to many DC-coupled audio amplifiers, except for the addition of controls peculiar to oscilloscopes (see Figure 9.10).

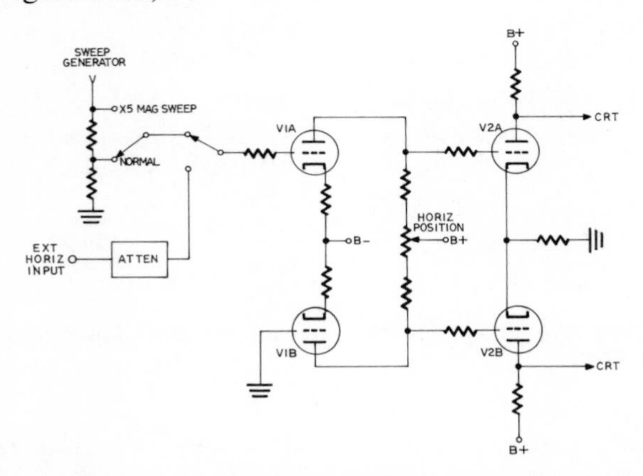

Figure 9.10 The horizontal amplifier amplifies the sweep sawtooth, and applies it to the horizontal deflection plates. Sometimes an external input signal replaces the sweep signal.

The *horizontal position control* is a potentiometer that changes the DC level of the amplifier so as to alter the voltage balance at the output, thereby shifting the display horizontally.

The *magnifier control* provided in many oscilloscopes enables you to increase the scale of the sweep by a factor of 5, 10, 100, etc., the factor varying with the oscilloscope model. The arrangement shown in Figure 9.10 gives you a choice of NORMAL or X5. In the NORMAL position, the sweep signal is attenuated by a factor of 5, but when the switch is moved to X5 this attenuation is bypassed.

The *external horizontal input* connector on the front panel of the oscilloscope is the means whereby an external signal can be fed to the horizontal amplifier. This is selected by means of an "external" setting on the sweep timing switch, or in some cases by a separate switch. In either case, the sweep generator is disconnected. An attenuator is also provided to adjust the input signal amplitude.

Vertical Amplifier

Most vertical amplifiers are differential amplifiers. As we saw in Chapter 1, differential amplifiers are used in test equipment because of their ability to reject common-mode signals. Figure 9.11 shows a simplified version (with attenuator and sync takeoff omitted). If the shorting strap is removed, the amplifier has a balanced input. Identical signals at terminals 1 and 2 will pass along identical paths to cancel each other out. If they are not identical, the difference will be amplified and appear at the output. This is a "floating" input, since neither side is grounded. Connecting the shorting strap between terminals 2 and 3 changes the balanced input to an unbalanced input by grounding terminal 2. V1B now acts as a phase-splitter to convert the single-ended input to a double-ended one, and the amplifier behaves like any other pushpull amplifier.

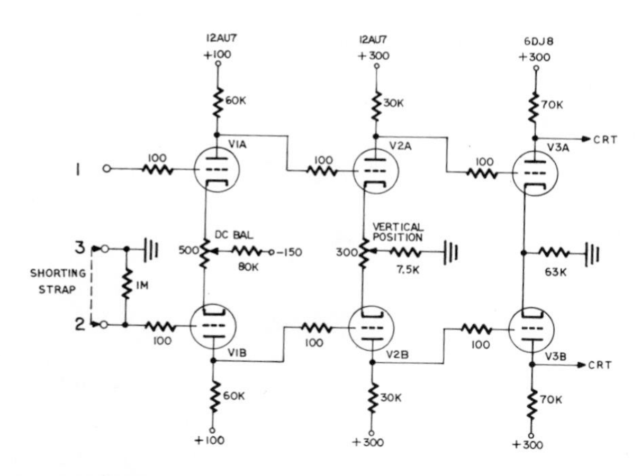

Figure 9.11 The vertical amplifier is a differential amplifier unless input terminal 2 is grounded to 3, when it becomes a push-pull amplifier. Its output is connected to the vertical deflection plates of the CRT.

The *DC balance control* is provided to balance the DC levels of each channel so that no vertical shift occurs on switching ranges on the vertical attenuator.

The *vertical position control* allows you to move the display up or down on the CRT screen by changing the bias on V2A and V2B in opposite directions, forcing one plate to a higher potential than the other.

Vertical amplifiers may also be considered as DC-coupled video amplifiers because of their wide bandwidth. For instance, the oscilloscope illustrated in Figure 9.1 has a frequency response from DC to 10 megahertz, which is not exceptional. There are instruments that go over a gigahertz. Obviously, such amplifiers must be very carefully designed. They all use negative feedback to obtain a flat response over as wide a range as possible, with peaking coils to provide high-frequency compensation.

Oscilloscope Probes

The input impedance of the vertical amplifier is equivalent to a high resistance shunted by a small capacitance, usually one megohm and less than 50 picofarads. However, even these values may produce undesirable loading in some circuits, so that its waveforms are not correctly reproduced on the CRT screen.

An oscilloscope probe contains a resistor shunted by a capacitor. The probe resistor is in series with the input resistance of the oscilloscope (Figure 9.12), so the combined resistance of both has a reduced loading effect on the circuit under test. For instance, if the probe resistor has a value of 10 megohms and the oscilloscope's input resistance is 1 megohm, the circuit sees a load of 11 megohms instead of the 1 megohm without the probe.

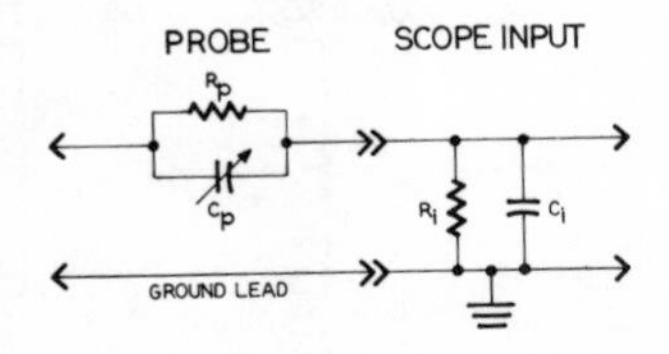

Figure 9.12 The Input Circuit of an Oscilloscope where a Probe is used

The probe resistor bears a fixed ratio to the oscilloscope input resistance. Together they form a voltage divider that attenuates the signal by a discrete amount: X1, X10, X100 or X1000. If your probe is marked X10, for example, the signal voltage will be divided by this factor. You must remember to multiply the voltage indicated on the CRT screen by ten to allow for this.

You won't need the probe capacitor if you make low-frequency measurements only, but the input capacitance of the oscilloscope has a shunting effect on the higher frequencies, which gets progressively worse as the frequency increases. The probe capacitor compensates for this by bypassing more of the higher frequencies around the probe resistor.

To adjust this capacitor, you connect the probe to a square-wave source, and see what it does to the square wave. A square wave may be thought of as consisting of a fundamental wave and an infinite series of odd harmonics, as in Figure 9.13. Notice how the first few harmonics combine with the fundamental to approximate a square wave. As higher-frequency harmonics are added, the leading edge of the complex waveform gets steeper and its corner sharper. If you could really include all the odd harmonics out to infinity, you would get a perfectly rectangular corner. While this is physically impossible—since there is no such thing as an infinite frequency—you can get corners that look pretty square. They are produced by the higher-frequency components of the square wave. On the other hand, the middle part of the flat top of a square wave owes its shape to the lower-frequency components.

Illustrating the addition of successively higher-order harmonics to a fundamental sine wave, to produce a close approximation of a square wave.

Waveform A is the fundamental sine wave. Waveform B shows the result of adding the third harmonic sine wave to the fundamental. Waveform C shows the effect of adding both the third and the fifth harmonics to the fundamental. Note that waveforms B and C show closer and closer approximations to the final square wave D.

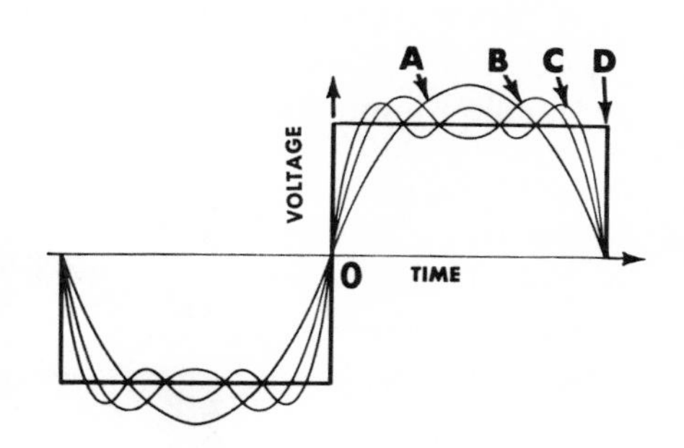

Figure 9.13 *(Reprinted by Permission of Tektronix, Inc.)*

This is why a square wave or rectangular pulse makes such a good test signal. You can see, by how it is distorted, what the circuit is doing to the various frequencies. If it retains its shape, all of them are being passed through without change of phase or amplitude. If the corners are rounded, the higher frequencies are being attenuated. If the top is not flat, the lower frequencies are affected.

If C_p is adjusted to bypass fewer of the higher-frequency components around R_p, the effect on the square wave will be as shown in Figure 9.14A; if more, as in B. You must adjust your probe until you get the optimum frequency response shown at C. This adjustment should be checked regularly, especially if you use this probe with a different oscilloscope or plug-in.

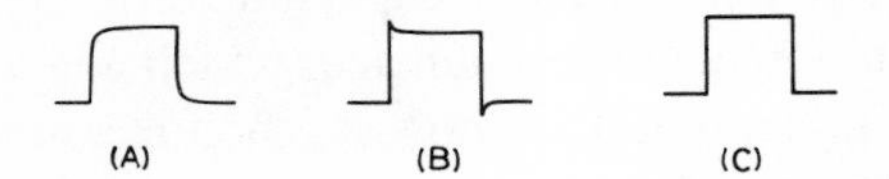

Figure 9.14 (A) High-frequency attenuation.
(B) High-frequency accentuation.
(C) Optimum high-frequency response.

OPERATING THE OSCILLOSCOPE AND INTERPRETING THE DISPLAY

Now let's take an actual oscilloscope, such as that in Figure 9.15, and see what it does. You begin by plugging the power cord into an AC outlet and turning on the power switch, which is incorporated in the BRILLIANCE control (top right, front panel). As this knob also controls the brightness of the display, do not turn it up too far, as too bright a trace can burn the phosphor screen of the CRT. You'll know if the power is on because the graticule will light up. (You can adjust the intensity of the graticule illumination by means of a knob in the rear.)

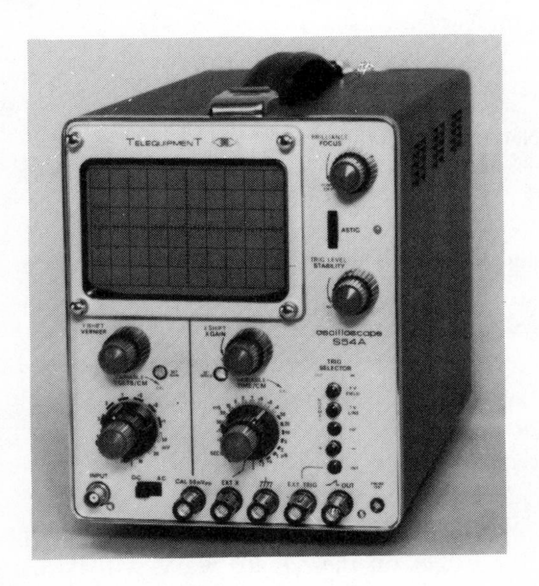

Figure 9.15 To conserve space on the front panel, concentric controls are used in many oscilloscopes. The smaller knobs are colored differently from the larger ones, and their names are in matching colors on the panel. *(Courtesy Tektronix, Inc.)*

Set the Y SHIFT and VERNIER, X SHIFT and STABILITY controls to midrange, and the TRIG LEVEL to AUTO. Rotate both VARIABLE controls fully clockwise to their CAL positions (this switches them off). Under TRIG SELECTOR depress the INT push button.

Now take a short coaxial test lead with a BNC connector at one end and a banana plug at the other, and connect it between the INPUT connector and the CAL 50 mVpp jack. Set the VOLTS/CM control to 10 mV, the TIME/CM control to 2 mS, the X GAIN to CAL, and the DC-AC switch to AC.

Adjust the BRILLIANCE, FOCUS and ASTIG (astigmatism) controls for a good picture. As mentioned above, the BRILLIANCE control should be turned up to give a bright, but not overbright, display. The FOCUS control narrows the electron beam, the ASTIG control rounds it. Together, these controls give you a display equally sharp in all directions.

You should now have a display similar to Figure 9.16. Use the Y SHIFT and VERNIER and X SHIFT controls to center it in relation to the graticule, as shown. You are looking at one cycle of the oscilloscope's internal calibrator signal of 50 millivolts peak-to-peak at 60 hertz. (Oscilloscopes respond to peak-to-peak voltages. The displayed value must be multiplied by .3535 to obtain the RMS value.) Check the Y and X measurements with the graticule lines. (These are at one-centimeter intervals; the small divisions on the vertical and horizontal center-lines are spaced at two millimeters.) The Y dimension should be 5 cm. The signal, therefore, has an amplitude of 5 x 10 mV (the VOLTS/CM setting), or 50 mV. If it does not, check if the VARIABLE knob is in the CAL position, and then with a small screwdriver adjust the SET GAIN control to set the amplitude of the display to 5 cm.

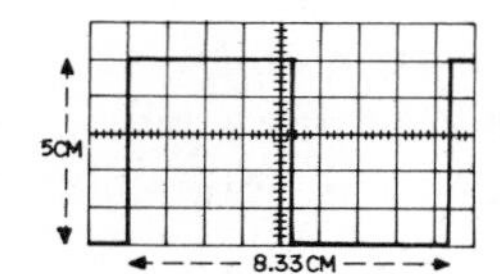

Figure 9.16 One cycle of the calibrator signal of the oscilloscope in Figure 9.15.

Now check the X dimension. This should be 8.33 cm. The time interval between the leading edge of the first square wave cycle and the next is, therefore, 8.33 x 2 mS (the TIME/CM setting), or 16.7 milliseconds. This is the time of one cycle of the calibrator signal. Its reciprocal is its frequency, or 60 hertz (1/.0167 = 60). If the X dimension is not 8.33 cm, check if the VARIABLE and X GAIN controls are in their

CAL positions, then with a small screwdriver set the X dimension to 8.33 cm by adjusting the SET SPEED control.

The accuracy of these settings depends on the accuracy of the voltage calibrator, of course. In this instrument, it is 50 mV ± 1 mV peak-to-peak, while its frequency is 60 hertz, ±0.1 hertz if on standard US line voltage. (This model can operate on line voltages with frequencies between 48 and 440 hertz.)

There will also be some slight variation between the calibrated steps of the VOLTS/CM and TIME/CM controls, due to the tolerances of the components used; so although you have set one step on each control to agree with the voltage calibrator, the others will have an additional tolerance of ±5 percent. Theoretically, therefore, it is possible for a voltage or frequency measurement to be as much as 7 percent off; statistically, however, an accuracy of ±4 percent is more realistic.

Now rotate the vertical VARIABLE control clockwise. The height of the display decreases. This control provides overlapping variable (and uncalibrated) attenuation between the calibrated steps of the VOLTS/CM switch, and extends the attenuation range to about 125 V/cm.

Try the same with the horizontal VARIABLE control. It works similarly, and extends the timing range to approximately 5 seconds per centimeter.

Return both VARIABLES to their CAL positions, and rotate the X GAIN control clockwise. It expands the size of the display laterally until it is approximately five times greater than when the X GAIN knob is all the way clockwise. This control increases the maximum sweep speed to approximately 40 nanoseconds per centimeter. Reset the X GAIN control to its CAL position.

When the TRIG LEVEL control is in AUTO, a sweep appears on the CRT screen without application of any type of synchronizing signal. However, when a synchronizing signal with a frequency between 50 hertz and one megahertz is applied, the sweep automatically triggers from this signal. The level at which it triggers is the average DC level of the signal. If the signal frequency is outside those limits, you should not use the AUTO position.

Rotating the TRIG LEVEL control when not in AUTO lets you set the triggering level at any position you select on the positive or negative slope of a signal, according to whether the TRIG SELECTOR polarity push button is set to + or −.

When the TRIG LEVEL control is not in AUTO, you may need to adjust the STABILITY control. Rotate it clockwise until a trace appears on the CRT, then back off a little from the point where the trace appears. The sweep should then trigger satisfactorily when a signal is applied.

What you have been doing with the calibrator signal is no different from what you do when measuring any other waveform. You connect the oscilloscope input to the proper circuit point, and adjust the controls to get a suitable display. Many schematic diagrams have representations of the waveforms that should be at various points in the circuit, together with their amplitude and frequency, and sometimes the oscilloscope vertical and horizontal settings also. Comparison of what you see on the CRT screen and what is shown on the schematic can tell you a great deal about the behavior of the circuit.

Where the circuit does not have any characteristic waveforms of its own, an external signal from a signal generator must be applied. A square wave is generally preferable for the reasons given in the discussion on oscilloscope probes above.

In the next chapter, you'll read more about basic measurements with oscilloscopes using plug-ins.

10

TROUBLESHOOTING WITH MULTIPURPOSE AND DEDICATED OSCILLOSCOPES

A conventional oscilloscope has all its circuits included in the single instrument. Compared to the plug-in type described below, it is smaller and less expensive. It can be made portable and battery-operated. Because it is self-contained it is also less flexible, and is normally used for voltage-time measurements only. Since it is designed for one measurement area, it is called a *dedicated oscilloscope*.

A plug-in oscilloscope is much heavier, more costly and the most versatile. As the name implies, some of its circuits are in the form of removable plug-in units. At one extreme, are models that contain only power supplies, CRT and associated circuitry in a *mainframe*. At the other, are models with all circuits built in except for a vertical preamplifier. Whatever the arrangement, the use of plug-ins makes it a *multipurpose oscilloscope,* adaptable for different uses by changing the plug-ins.

MULTIPURPOSE OSCILLOSCOPES

Figure 10.1 illustrates a multipurpose oscilloscope with four plug-ins. By using all the plug-ins available for this instrument it is possible to have more than twenty different operating modes in one oscilloscope. Another interesting feature is the CRT readout appearing above the waveforms, whereby the principal parameters of each channel are displayed in numerical form. This makes it easier to read the instrument and avoid making errors such as overlooking probe attenuation and amplifier gain factors.

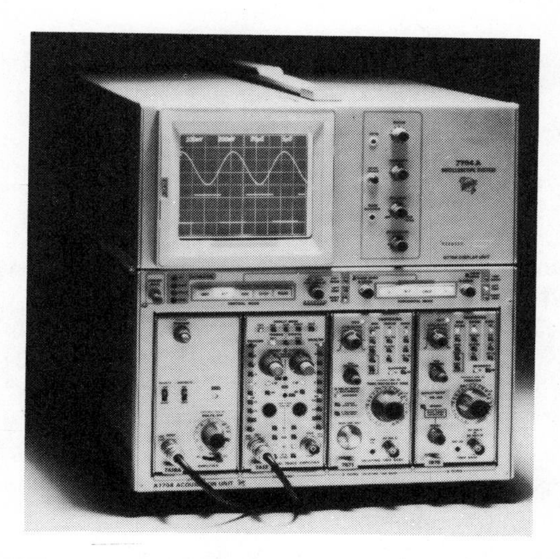

Figure 10.1 A multipurpose oscilloscope with a mainframe containing the CRT and its controls and circuits, the electronic switching circuits, calibrator and power supplies; and two vertical and two horizontal plug-ins. The readout above the waveforms gives the vertical sensitivity and time-base control settings of the plug-ins, going from left to right. A majority of the controls are push buttons that light when activated. *(Courtesy Tektronix, Inc.)*

The basic instrument has two vertical channels and two horizontal channels (see Figure 10.2). Electronic switching makes it possible to use all four at once to display two entirely different signals simultaneously. Switching is done in the mainframe. The two electronic switches operate together to switch the CRT alternately to the left and right vertical and horizontal channels. The switching is done at high speed, so the eye sees two apparently separate displays on the CRT screen because it cannot follow the alternations.

There are two ways of sharing the CRT between the left and right channels, called *alternate* and *chopped*. In the alternate mode, the elctronic switch operates as a bistable multivibrator, remaining in one state during all of each sweep, and switching to its other state on receipt of a sync pulse from the sweep generator at the end of the sweep. In other words, the display consists of alternate sweeps from each horizontal plug-in, coupled with the corresponding vertical plug-in signal. In the meantime, the other sweep and vertical signal are cut off.

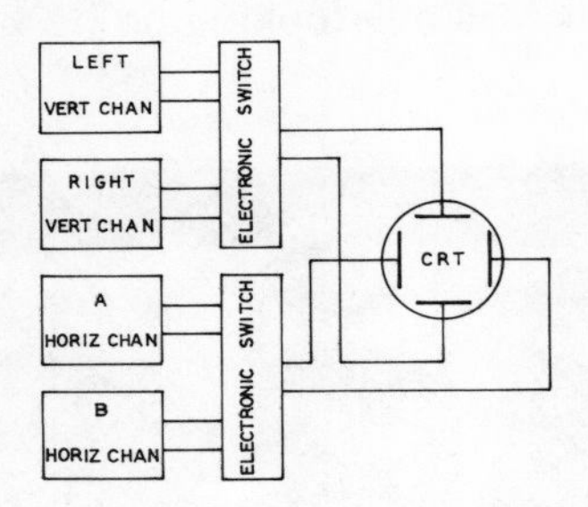

Figure 10.2 Block diagram to show how the four plug-ins in the oscilloscope in Figure 10.1 are connected to the CRT.

In the chopped mode, the electronic switch runs as a synchronized astable multivibrator, alternately turning the channels on and off at a fixed rate of 100 kilohertz or one megahertz (selectable internally).

The chopped mode is better for lower-frequency input signals, the alternate for higher frequencies.

Each vertical plug-in has its own vertical-positioning control, so the two waveforms can be moved up and down on the screen independently of each other, and may be displayed separately or superimposed.

Plug-Ins

The principal types of plug-ins used in multipurpose oscilloscopes are:

Vertical plug-ins	Single-channel amplifier Dual-channel amplifier Differential amplifier
Horizontal plug-ins	Time base Delaying time base

There are also many specialized plug-ins that enable the instrument to be used for other functions, such as curve-tracing, spectrum analysis, sampling, time-domain reflectometry, and so on, that will be discussed later.

SINGLE-CHANNEL AMPLIFIER PLUG-IN

A single-channel amplifier plug-in, similar to that occupying the left-hand position in Figure 10.1, consists of a wide-band amplifier with a single-ended input and a push-pull output. The input is via the BNC connector shown, and has an impedance of one megohm shunted by 20 picofarads. A three-position switch to the right of the GND position allows you to select DC (straight through) or AC (via a capacitor) connection to the VOLTS/DIV attenuator. The GND position grounds the input when you want to set the trace to a zero reference (the bottom line of the CRT graticule, for example). Then you can go back to the DC position to see both AC and DC components of the signal, or to the AC to block the DC component, if any.

The VOLTS/DIV step attenuator is similar in principle to others already described, and has a VARIABLE control also. Slide switches select the polarity of the display and the bandwidth, which can be either FULL (170 megahertz) or 20 MHZ for lower-frequency signals.

Two of these plug-ins may be used for the two vertical channels, as we've seen already, or one may be used in a vertical channel with another in a horizontal channel for *X-Y operation*.

X-Y DISPLAYS

In X-Y operation you apply a second signal to the CRT instead of the sawtooth from the sweep generator. This gives a display that is a graph of the functional relationship between the two signals (the vertical and the horizontal). This mode of operation is very useful for comparing the phase angle of two signals at the same frequency. It can also be used to determine frequency ratio and amplitude ratio.

If you adjust the vertical and horizontal input attenuators for equal amplitude signals at their deflection plates, the display will be in the form of one of the *lissajous patterns* shown in Figure 10.3. The zero phase angle display is often used to set a synchro or other phase-sensitive device to zero degrees. The device under test is adjusted until the elliptical display closes up to a straight line, when the phase angle must be either zero or 180 degrees, according to the slope.

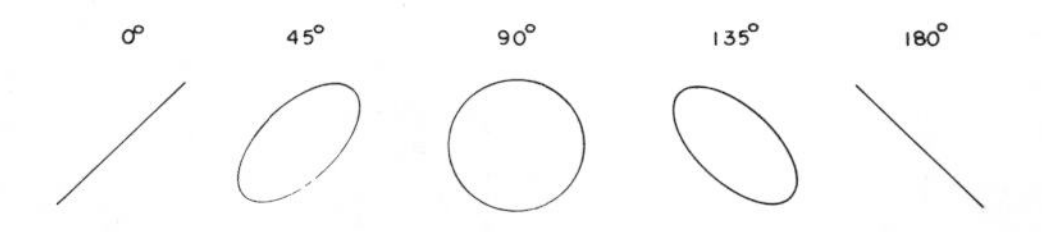

Figure 10.3 Lissajous patterns show the phase relationship between two signals of the same frequency when one is applied to the vertical input and the other to the horizontal input, with the sweep generator disconnected.

You do not need a plug-in oscilloscope to do this, but for really accurate results the vertical and horizontal amplifiers must be identical, and the instrument must not have a delay line, if spurious phase shift is not to be added to the real phase shift.

DUAL-TRACE AMPLIFIER PLUG-IN

This plug-in contains two separate channels (CH1 and CH2) with an electronic switch that can give an alternate or chopped display as described above for two single-channel plug-ins. If one of these plug-ins is used for each vertical channel of the oscilloscope, you can get four separate displays on the CRT screen at the same time. (In Figure 10.1, a single-channel plug-in is combined with a dual-trace plug-in.)

Five push buttons across the top of the plug-in front panel select CH1, ALT, ADD, CHOP, or CH2. ADD is a mode in which the two input signals can be added algebraically. This is very useful when you have an undesired signal with the one you want to view. You connect both signals to CH1 input and the undesired signal to CH2. Then you depress the CH2 INV (invert) push button, which reverses the polarity of the CH2 input (changes it from positive to negative, or vice versa). The unwanted signal is now largely canceled, although you will have to do some minor adjusting of the CH2 attenuator.

Talking about attenuators, you'll see that the step attenuators for both channels use lighted push buttons instead of rotary switches. This is because of space limitations (the panel is only 2⅝ by 5 inches); push-buttons are used for all controls except the variable ones.

In Chapter 9, you saw how to measure the time base and AC-voltage components of a waveform. The procedure using a dual-trace plug-in is exactly the same when you select either channel and apply a signal to its input.

Measuring the DC level of a signal is similar, except that you first set the input selector switches to GND, and the triggering controls for a free-running sweep. You then position the trace along one of the horizontal graticule lines as in Figure 10.4, where it is set along the bottom line at

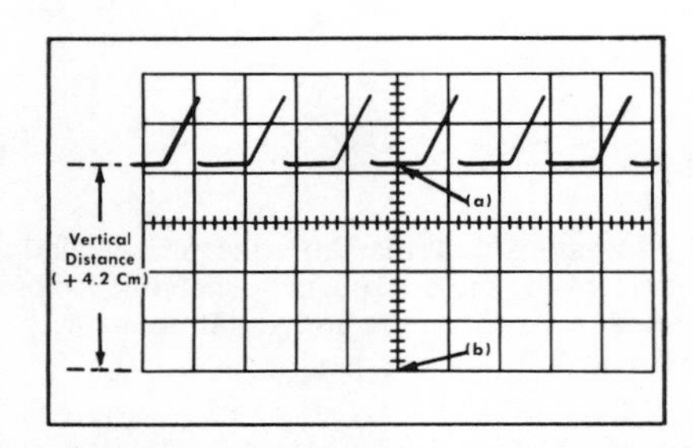

Figure 10.4 Measuring the DC level of a signal. *(Reprinted by permission of Tektronix, Inc.)*

(b). This line is now the ground or zero-volts reference line. (For a negative DC component you would use the top line.) Once you have set the reference do not touch the POSITION control again.

Now set the input selector to DC, and connect the signal to the input. Adjust the controls for a stable display, and measure the vertical distance in centimeters from the reference line (b) to the DC level at (a). Multiply this by the VOLTS/CM switch setting and the probe attenuation factor, if any. For example:

Vertical Deflection		*VOLTS/CM Setting*		*Probe Attenuator*		*DC Component*
4.2 cm	X	2V	X	10	=	84V

This particular plug-in also has an OFFSET function. This consists of an internal DC source, continuously variable between +1V and −1V according to the control setting, which may be used to offset the trace (see further details under Differential Amplifier Plug-In).

A dual-trace plug-in enables you to measure the phase-difference between two sine waves of the same frequency. You apply one signal to the CH1 input, the other to the CH2, and set the display mode to ALT or CHOP. As triggering may be difficult (because of the phase difference), use external triggering from the leading signal, unless the plug-in has a switch for disabling internal trigger signals from one of the channels, as some do. Adjust the sweep rate to obtain a display like that in Figure 10.5, and set the VOLTS/CM and VARIABLE plug-in controls so both waveforms are equal and fill the graticule area vertically.

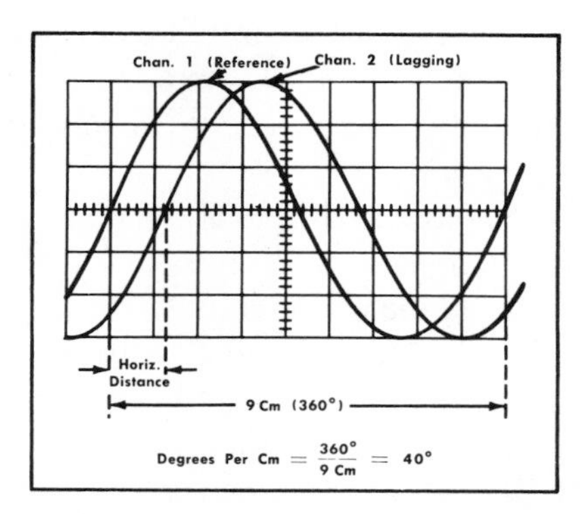

Figure 10.5 Measuring the phase difference between two signals of the same frequency. *(Reprinted by permission of Tektronix, Inc.)*

Now adjust the horizontal VARIABLE sweep control until one cycle of the reference signal occupies 9 centimeters horizontally. This means that each centimeter of the graticule equals 40 degrees, and each small division is 8 degrees. The horizontal distance in centimeters, multiplied by 40, is the phase difference. If the angle is small, increase the sweep rate (do not touch the VARIABLE control) or use the magnifier to get a display like that in Figure 10.6, but remember that your multiplying factor will be less. For instance, if you make the sweep rate five times as fast, each graticule centimeter will now be eight degrees instead of forty.

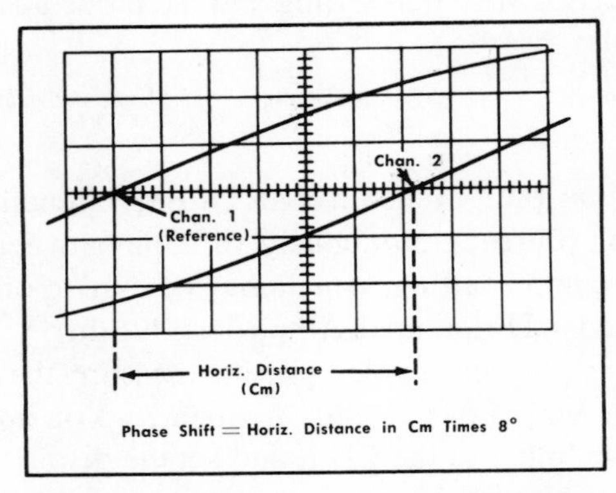

Figure 10.6 Measuring a smaller phase difference than in Figure 10.5. *(Reprinted by permission of Tektronix, Inc.)*

DIFFERENTIAL AMPLIFIER PLUG-IN

A differential amplifier plug-in also has two inputs, but they do not apply the signal to different channels as in a dual-trace amplifier plug-in. There is one push-pull amplifier that amplifies the *difference* between the input signals. If there is no difference, there will be no output. If a signal is applied to one input only, that signal will be amplified in the same way as in a single-trace plug-in.

As we mentioned before, a differential amplifier is able to cancel out an unwanted signal if it applied to both inputs. Such a signal is called a common-mode signal, and the cancellation is called common-mode rejection. Since some of the unwanted signal will get through anyway, the ratio of the amplitude of the common-mode input signal to the amplitude of the difference signal displayed on the CRT is called the *common-mode rejection ratio* (CMRR), which should always be several thousand to one.

A type of differential amplifier plug-in is a *differential comparator,* which contains an accurately-calibrated adjustable voltage source. This voltage is applied differentially to *offset* any unwanted portion of the applied signal, and allows you to measure relatively small AC or DC signals riding on relatively large ones. Differential comparators can make the following measurements in addition to the ordinary ones:

A. DC voltage

B. Small AC or DC signals superimposed on DC

C. Small AC signal variations on large AC signals

D. High-amplitude low-frequency AC

TIME-BASE AND DELAYING TIME-BASE PLUG-INS

A time-base plug-in contains the circuits described in Chapter 9 under Sweep Triggering and Synchronizing, plus a horizontal amplifier, so we need not to go over them again.

A delaying time-base plug-in, when used with a time-base plug-in, has the additional capability of expanding a selected portion of the display. In Figure 10.1, the left-hand horizontal plug-in is a delaying time-base plug-in. The most noticeable difference between it and the time-base plug-in to the right is the DELAY TIME MULT (multiplier) control in the bottom left-hand corner.

This control sets the threshold level of a comparator in the delaying time-base plug-in. When the sawtooth sweep voltage in this plug-in reaches the threshold level, the comparator generates a trigger pulse that starts the sweep in the other time-base plug-in.

In oscilloscopes in which the horizontal deflection circuits are part of the mainframe, delaying sweep is often provided also. In these instruments, two time bases labeled A and B are included. The horizontal display selector control can be set to use A or B independently or in combination. In the combination "B intensified by A," the display will look like the upper part of Figure 10.7. Here the A sweep is superimposed on the B sweep, causing a portion to be brightened or intensified. As the A-sweep speed in this case was set to ten times the B-sweep speed, the brightened portion is only one-tenth of the B-sweep trace. Its length can be varied by changing the sweep rate.

The delay-time multiplier control adjusts the time interval between the start of the B sweep and the start of the A sweep. Rotating this control causes the brightened portion to move back and forth on the trace, depending on the direction of rotation. This enables you to position it so that it intensifies any part of the trace you are especially interested in. Then you switch the horizontal display control to "A delayed" and the brightened portion expands to the full width of the graticule, as in the lower part of Figure 10.7.

This is somewhat similar to a magnified sweep, but with greater ratio of effective magnification and better resolution and stability.

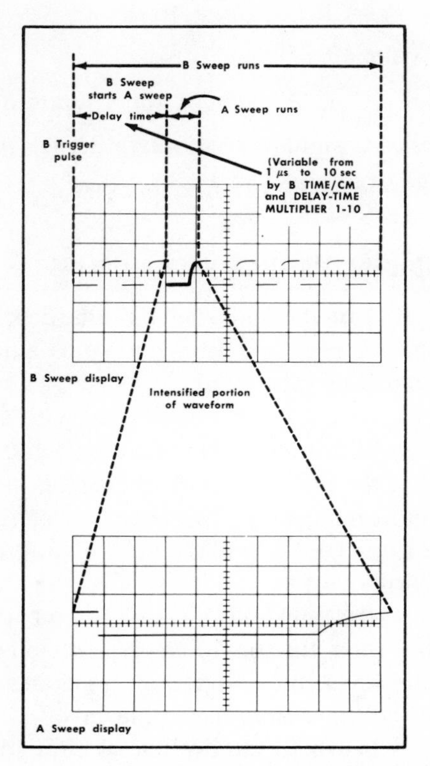

Figure 10.7 How delaying sweep works. *(Reprinted by permission of Tektronix, Inc.)*

The delay-time multiplier control can also be used to compare the time interval between two positions on the B sweep. This is done by rotating the DELAY TIME MULT control to locate the start of the brightened portion at the first position, noting the multiplier dial reading, then locating it at the second position and taking a second reading. The difference between the two readings, multiplied by the B sweep time, gives the time interval between the two positions.

SOME OTHER PLUG-INS

Most of the plug-ins in general use are variations of the types described above. There also are some other more specialized plug-ins that convert the multipurpose oscilloscope for special uses. Such special uses are also provided for by corresponding dedicated oscilloscopes.

The curve-tracer plug is used to display the characteristic curves of small-signal semiconductors. It converts a multipurpose oscilloscope to a curve-tracer oscilloscope as described previously in Chapter 5.

Whether plug-in or dedicated oscilloscope, a curve-tracer is capable of displaying one or more characteristic curves of two-and three-terminal devices and integrated circuits (ICs). Each curve is developed by driving one terminal with a constant voltage or current, then sweeping the other with a half sine wave of voltage. If more than one curve is to be displayed, the driving source is stepped through several values, and the sweep repeated once for each step, giving a display like that in Figure 5.8.

Characteristic curves of semiconductors are used to sort components, predict performance in a circuit, check devices against their specifications or design better semiconductors. Characteristics of practical importance to their use in a circuit can usually be measured best with a curve-tracer.

In addition to the curves, some curve-tracers also give a digital readout on or beside the CRT screen of parameters such as β or g_m per graticule division.

A *time-domain reflectometry* (TDR) plug-in is basically like radar. It is used to locate trouble spots in transmission lines. The TDR plug-in contains a pulse generator that sends a pulse along the line. This pulse is also applied to the vertical channel of the oscilloscope, triggering a sweep. If a fault exists in the transmission line, voltage is reflected back from it to appear as a blip on the trace. The horizontal distance from the beginning of the trace to this blip shows the time it took the pulse to travel to the fault, and for the reflection to come back.

Each type of transmission line transmits signals at a characteristic speed, so the distance along the line to the fault can be calculated easily. However, the plug-in saves you the trouble by providing a direct-reading dial on which you can read the time or distance in appropriate units. Different scales are provided for air or polyethylene dielectric lines.

A *spectrum-analyzer* plug-in enables the oscilloscope to display the energy distribution of an electrical signal as a function of the frequencies within that signal. This plug-in is a swept receiver giving a display of amplitude (vertical) against frequency (horizontal). This differs from the usual display of amplitude versus time, where the view is of a train of events that follow one after the other. Here you are looking at a cross-section of the waveform at an instantaneous moment, showing its center frequency and sidebands. This plug-in can also be used for checking frequency response of filters and measuring harmonic distortion and modulation.

A *sampling* plug-in employs a technique very similar in principle to the use of a strobe light to study fast mechanical motion. As each cycle of a fast repetitive signal comes in to the plug-in, it takes a small sample of it and displays it on the CRT screen. This sample is only a dot, but its vertical position corresponds to the instantaneous voltage of the signal at that moment.

Each sample is taken from a successive cycle at a point a little bit further along the signal waveform than its predecessor. Because of the high speed of the action, a complete series of dots appears seemingly simultaneously on the CRT screen. Although each is taken from a separate cycle, they are closed up on the screen so that they trace out one cycle, as shown in Figure 10.8.

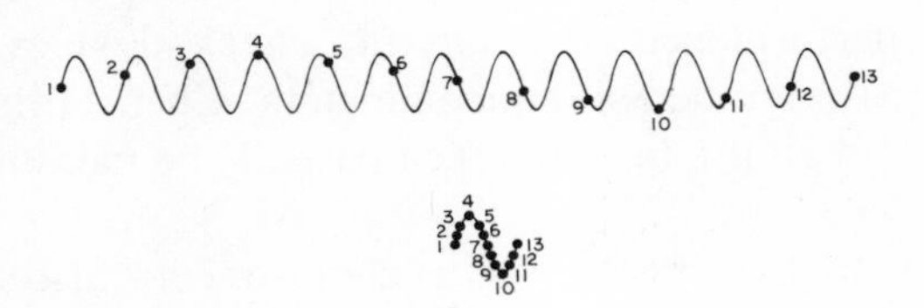

Figure 10.8 A sampling plug-in takes small samples of successive cycles of a high-frequency signal, and combines them to give a one-cycle representation of the waveform, as shown above. Of course, in the plug-in a much larger number of samples is taken, so the waveform has a smoother appearance.

Other plug-ins include *digital multimeter* plug-ins and *counter-timer* plug-ins that convert the oscilloscope into a digital multimeter or electronic counter (with digital readout on the CRT screen); also *current amplifier* plug-ins (with special clip-on probes similar in principle to those described in Chapter 6), and plug-ins for use with storage oscilloscopes. From this you can see how versatile a multipurpose oscilloscope can be.

DEDICATED OSCILLOSCOPES

There is little difference in performance between a multipurpose oscilloscope with a curve-tracer plug-in and a curve-tracer oscilloscope. In fact, the performance of the curve-tracer oscilloscope may be somewhat better, since it is dedicated to this particular function. The chief advantage of the multipurpose instrument is its versatility, but it is more expensive and heavier than the dedicated oscilloscope.

For this reason, all portable oscilloscopes can be described as dedicated, general-purpose oscilloscopes without plug-ins. They are pretty much like the simple oscilloscope described in Chapter 9, which weighs only 17 pounds. This is an average weight for portables, which range from some 36 pounds down to three pounds (see Figure 10.9).

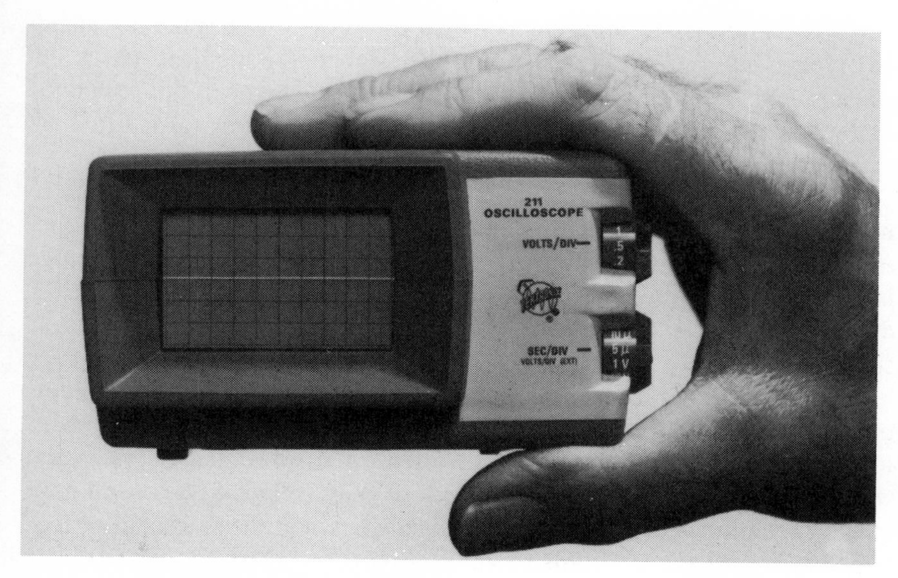

Figure 10.9 A really portable oscilloscope that weighs only three pounds! The controls are at the side. *(Courtesy Tektronix, Inc.)*

Many portables are also battery powered, using rechargeable nickel-cadmium battery cells. These will allow several hours operation between charging, but add somewhat to the weight.

All modern portables are solid-state instruments (except for the CRT), capable of withstanding shock, vibration and considerable environmental changes.

Apart from dedicated instruments that duplicate the plug-in functions already described, there are also oscilloscopes, both plug-in and dedicated, that become special-purpose oscilloscopes by the use of transducers and other auxiliary equipment. For instance, an *engine analyzer* comprises an oscilloscope with pressure and vibration transducers, ignition pickoff, magnetic pickup and rotational function generator. Using a four-channel plug-in, this instrument can show simultaneously cylinder pressure, ignition timing and valves opening and closing (by vibration), against an array of crank-angle markers supplied by the rotational function generator (which is mechanically coupled to the engine under test). From this a technician can study the overall performance of the engine.

Another important area in which the oscilloscope plays a vital part is the medical field, where it forms the display element of various physiological monitors such as electrocardiograph (ECG) and electroencephalograph (EEG) equipment.

Oscilloscopes forming part of medical systems are generally standard types, but an instrument specially designed for medical use is the *physiological monitor*. It is portable, so that during an operation it may be mounted on the anesthesiologist's gas machine, and afterwards lifted off and carried with the patient to the recovery room.

Interface is achieved through patient cables and electrodes such as needles, plates or other types. The oscilloscope is provided with a beeper that gives an audible indication of the patient's heart activity. Any change in its sound will immediately alert the anesthesiologist to check the CRT display. If there is a complete loss of signal for any reason, an alarm is sounded.

Another group of oscilloscopes is designed for the television industry. These include the *waveform monitor,* which is a conventional oscilloscope with characteristics best suited to display television-signal waveforms; and the *vectorscope*, which displays relative phase and gain of the chrominance signal. Spectrum analyzers and time-domain reflectometers are also used in this field.

At the beginning of Chapter 9, we said that the oscilloscope was the most versatile of all types of test equipment. By this time you probably agree. As time goes by, more and more specialized plug-ins and dedicated instruments will be devised, enabling the oscilloscope to perform additional functions, or extending further the range of existing ones.

11

UNDERSTANDING MICROWAVE TEST EQUIPMENT

Microwaves have wavelengths measured in centimeters or less. This makes them quite different from signals of lower frequencies, whose wavelengths are measured in hundreds of meters. The instantaneous voltage of the latter will be the same everywhere in an average-sized circuit, but the instantaneous voltage of a microwave signal will be dependent on the point where you measure it.

The high frequency of a microwave signal would cause a radiation field to form around any ordinary conductor, from which some of the energy would be propagated as from an antenna. For such signals, therefore, leads must be practically non-existent or, where they have to be provided, transmission lines must be used. A transmission line is a conductor that behaves as the two halves of a dipole antenna would if they were placed close together and parallel to each other: their radiation fields would cancel each other out. Some transmission lines do, in fact, consist of two parallel wires as, for example, TV twin ribbon lead. However, coaxial cables or waveguides are the types mostly used, since their construction also provides shielding.

Microwave frequencies as presently defined start at one gigahertz. You can use coaxial components up to 18 gigahertz, but as you can also get waveguide components for frequencies starting at 0.320 gigahertz there is a considerable overlap. The following table shows the main waveguide frequency bands:

Band	Frequency	Band	Frequency
S band	2.6 - 3.95 GHz	X band	8.20 - 12.4 GHz
G or C band	3.95 - 5.85 GHz	P or Ku band	12.4 - 18.0 GHz
J band	5.30 - 8.20 GHz	K band	18.0 - 26.5 GHz
R or V band	26.5 - 40.0 GHz		

Every transmission line has its characteristic impedance (Zo), which is the ratio of voltage to current at every point along it, assuming there are no standing waves. There will be no standing waves if the line is of infinite length, or is terminated with its characteristic impedance so it looks to the source of the signal as if it *were* of infinite length. A terminating impedance of some other value will cause power to be reflected back toward the source. These reflected waves interfere with the direct waves, causing standing waves of voltage and current along the line (see Figure 11.1).

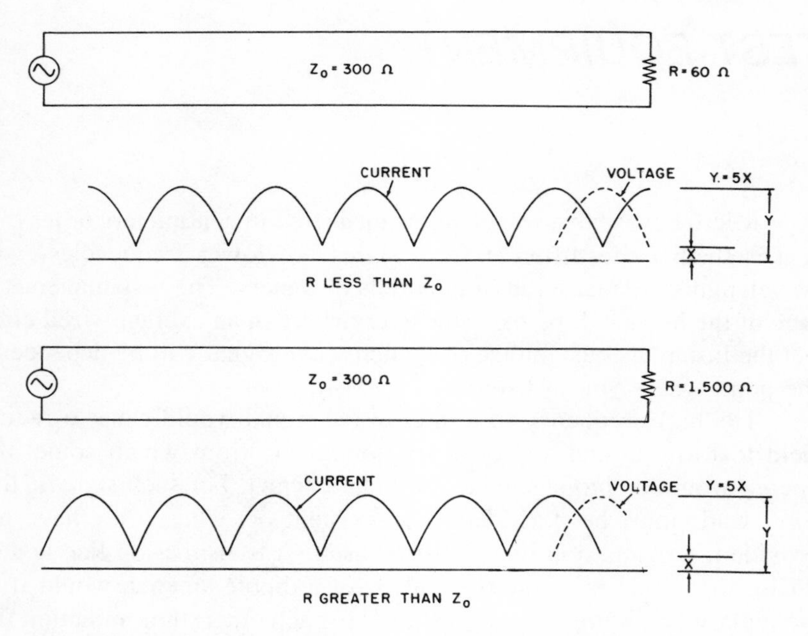

Figure 11.1 Mismatched lines showing standing-wave ratio

The peaks of the standing waves, where the voltage or current is at its maximum value, are called *loops*. The minima are called *nodes*. The ratio of the value at a loop to the value at a node is called the standing wave ratio (SWR). The SWR is also equal to the ratio of the characteristic impedance of the line to the impedance of the load. Because of the relationship between SWR, Zo and ZL, SWR measurement is used to determine the value of an unknown load impedance.

Impedance measurement is one of the three principal areas of measurement at microwave frequencies, the other two being measurement of attenuation and power. Attenuation measurements are the easiest to do, so we'll begin with them.

ATTENUATION MEASUREMENTS

The attenuation characteristic of a microwave component is called its *insertion loss,* which is the ratio of the power delivered to a matched load by a matched generator *before* and *after* the insertion of the component into the line or waveguide. Insertion loss consists of two elements: reflective and dissipative. The first is due to mismatch, since no component is ever perfectly matched in impedance. The second is due to power that would be dissipated within the component even if it were perfectly matched. The second element is really the true attenuation, but is inseperable from the first.

To obtain good accuracy in attenuation measurements you have to watch for trouble, and anticipate it, with:

Signal leakage

Impedance match

Generator protection

Equipment stability

Signal Leakage

Leakage can be into or out of the circuit. It is usually due to poor mechanical conditions, such as faulty assembly, defective parts, bent or dirty waveguide flanges. You can check for it by covering various points in the circuit with your hand, while watching the output indicator for a change in reading.

Impedance Match

Proper impedance matching requires the use of matching pads and tuners to obtain a better voltage standing-wave ratio (VSWR). Some crystal detectors change their impedance when the power level changes. Isolation by means of a matching pad is the simplest solution. Some others will be mentioned under impedance measurements.

Generator Protection

Isolation of the signal generator is always desirable, so that the frequency of the output signal will not be affected by changes in match as the device under test is inserted, or its attenuation is adjusted.

Equipment Stability

Obviously, accurate measurements cannot be made with equipment that varies its behavior while in use. Your power meter, SWR meter or signal generator must be stable and capable of performing according to its specification. Flexible cables and waveguides also have a nasty habit of changing their insertion loss when moved, so your setup should be such that this is not likely.

Measurement Methods

There are three common methods for making attenuation measurements:

Audio substitution

IF substitution

RF substitution

They are all basically similar, and may be used with single-frequency or swept techniques. We'll describe the audio substitution method here, using the setup shown in Figure 11.2.

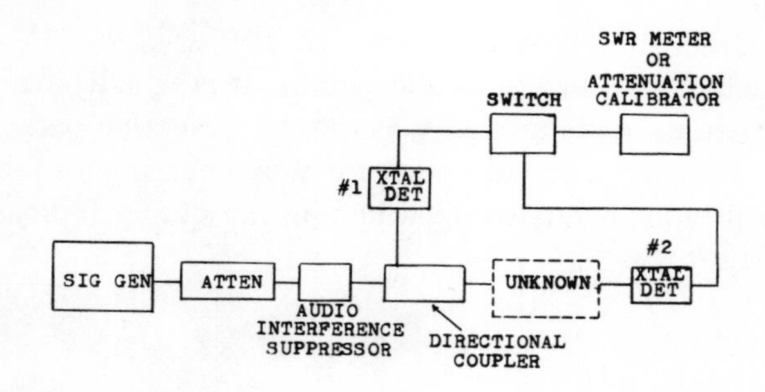

Figure 11.2 Test equipment setup for measuring attenuation by the audio substitution method

AUDIO SUBSTITUTION METHOD USING SINGLE FREQUENCY

In this method, you use an RF signal generator to produce a microwave carrier modulated with a 1-kilohertz audio signal. This signal passes through the attenuator being calibrated, and is then demodulated. The resultant audio signal is applied to an SWR meter with a built-in precision attenuator (sometimes an attenuator calibrator is used, working on the same principle). You then adjust the gain of the meter's amplifier to give a convenient meter reading.

The "unknown" attenuator is then removed from the circuit, and the detector is connected directly to the line. The precision attenuator is readjusted until the meter reads the same as it did when the attenuator was in the circuit. This means that you've put back into the circuit the same amount of attenuation you took out when you removed the "unknown" attenuator. The difference between the two settings on the precision attenuator must, therefore, be equal to the attenuation of the "unknown" attenuator.

Now let's look at the setup in Figure 11.2 in more detail. As explained already, the components and the waveguide or coaxial line connecting them must be well matched for impedance. The signal generator is protected by the 10-dB attenuator or pad. Next to it, in coaxial lines, it's good practice to use an audio interference suppressor. This is a small capacitance that blocks audio leakage from the generator, while passing RF with little or no attenuation.

The *directional coupler* allows a sample of the microwave signal to be taken out and applied to Detector #1, where it is demodulated. The resultant audio signal is then measured by the SWR meter, when the switch is in the proper position. This allows you to monitor the signal level to ensure that it is the same with or without the "unknown" attenuator in the circuit.

Since you'll use the various devices in this circuit for other purposes in microwave measurements also, this will be a good place to discuss how they work.

Microwave Attenuators

Microwave attenuators come in almost limitless combinations of physical and electrical characteristics. The most important of these are:

Peak power handling capability

Range of attenuation

Flatness with frequency

Physcial dimensions

Your selection of an attenuator starts from how it measures up to your requirements. As a general rule, you should not choose one that exceeds those requirements by very much, because you pay for what you get, and microwave equipment is expensive!

Most attenuators operate by having within them resistive material that absorbs energy. For this reason, they are called *dissipative attenuators*. The resistive material may be a thin metal film or aquadag (graphite), coating a glass rod or strip. This replaces the center conductor in a coxial line, or is placed parallel to the electrostatic field in a waveguide section.

In a fixed attenuator this resistive element doesn't move, but in variable attenuators a mechanical means is provided to alter the degree of attenuation. In waveguide attenuators this is done by making the resistive strip movable sideways, the attenuation increasing as it approaches the centerline of the waveguide. In coaxial types sliding or telescoping arrangements are used.

Non-dissipative attenuators do not absorb power. Instead, they couple more or less of the signal from the input to the output. In waveguide this may be done very simply by using the wrong size of waveguide for the frequency. In coaxial attenuators the coupling between coupling loops or lines can be adjusted to get the exact attenuation required.

DIRECTIONAL COUPLERS

The basic construction of a waveguide directional coupler is illustrated in Figure 11.3. It consists of a regular waveguide section, with another section attached in parallel with it. The energy from the source

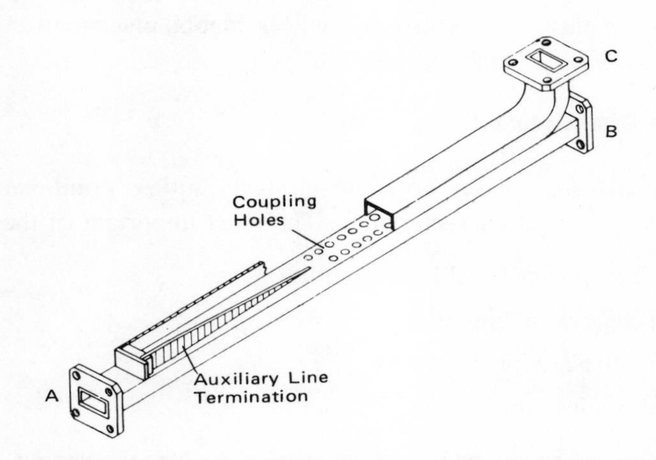

Figure 11.3 Cut-away view of a typical waveguide coupler, showing coupling holes and auxiliary line termination. Couplers are also available in four-part configurations, for use with external terminations.

enters at A, and most of it emerges at B. However, a sample of it passes from the main line to the auxiliary line, through the holes in the dividing wall. This energy comes out at C.

Only energy entering at A can be sampled. Reflected energy entering at B can pass through the holes, but is absorbed by the tapered resistive element, and disappears. This is why it is called a directional coupler—it couples energy coming in from one direction only.

Coaxial directional couplers work on the same principle, but have an arrangement of precisely-spaced bars to provide the coupling means, as shown in Figure 11.4.

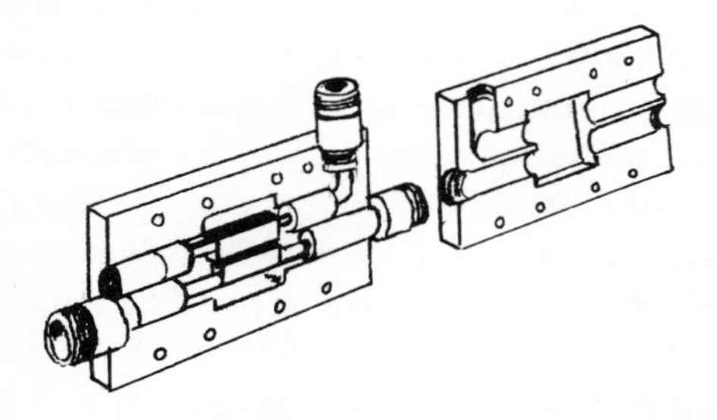

Figure 11.4 A coaxial directional coupler works on the same principal as a microwave coupler, but the coupling is done with precisely-spaced bars, as shown in this example with its cover removed.

The ratio, in decibels, of the forward power in the directional coupler main line to the power appearing at the auxiliary output is the *coupling coefficient*. Typically, this will be in values of 3, 10, 20 and 40 dB.

The ratio of the power coupled out in the preferred direction to that coupled out in the opposite direction is the *directivity*. There is always some, since nothing is perfect, but in a good waveguide coupler this ratio will be 35 to 40 dB over the frequency band the coupler is designed for. Coaxial types are usually a little less efficient.

Microwave Detectors

Microwave detectors consist of crystals, thermistors and barretters. We shall discuss thermistors and barretters in power measurement, for which they are mainly used. Detector crystals and diodes are the same in principle as those used for demodulation at lower frequencies. They are, therefore, square-law detectors, for which the output signal current strength is proportional to the square of the microwave input voltage.

When used for microwave detection, these detectors have to be capable of operating at the microwave frequency, and must be of the proper cartridge shape to fit in the *detector mount*.

The detector mount is either waveguide or caoxial, with an input that mates with the directional coupler's C output, or with the output of the unknown attenuator. A BNC connector is provided to couple out the audio signal.

SWR Meters

SWR meters are designed to measure the ratio between the maximum and minimum values of standing waves, and we shall see how they do this in the next section, dealing with impedance measurement. The meter consists of an amplifier with adjustable gain, and VTVM circuit to drive the meter, and a calibrated attenuator. Where the latter is very accurately designed, the SWR meter may be called an attenuation calibrator, but it is basically the same instrument. The amplifier is tuned to 1 kilohertz, the usual audio frequency, although SWR meters can be obtained for other frequencies. There is a provision for altering the tuning slightly, to allow for component-aging detuning. The meter dial is usually marked in SWR from 1 upwards, and in decibels. An example of an SWR meter is shown in Figure 11.5.

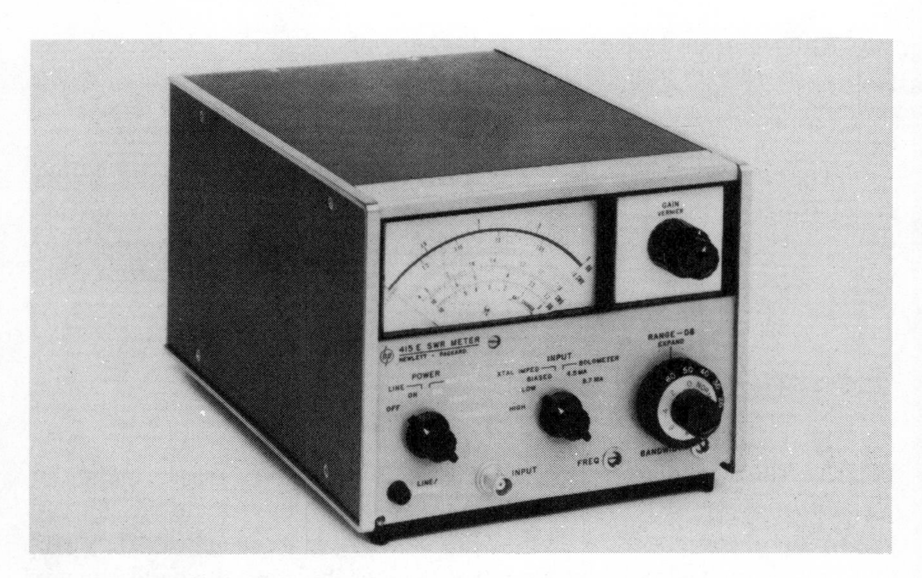

Figure 11.5 Hewlett-Packard Model 415E SWR Meter

SWEPT ATTENUATION METHOD

Figure 11.6 shows you how you can use a sweep generator and oscilloscope to calibrate an attenuator. The sweep generator (Figure 11.7) has an automatic level control (ALC). A sample of its output is fed back by the first directional coupler to a detector, and the latter's audio output is connected to the ALC input on the sweep generator to control the level of the sweeper's output signal. This keeps the output at the level at which you set it, regardless of changes in the external circuit from connecting and disconnecting the attenuation under test, as well as variable response to different frequencies.

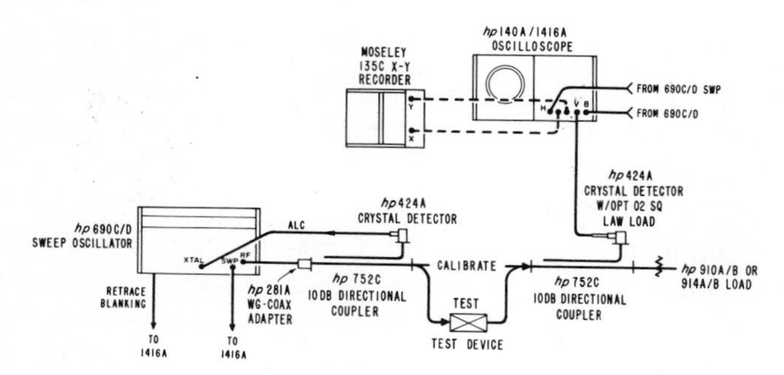

Figure 11.6 Swept attenuation system for measurements up to 40 db with oscilloscope or X-Y Recorder read-out. *(Courtesy Hewlett-Packard Co.)*

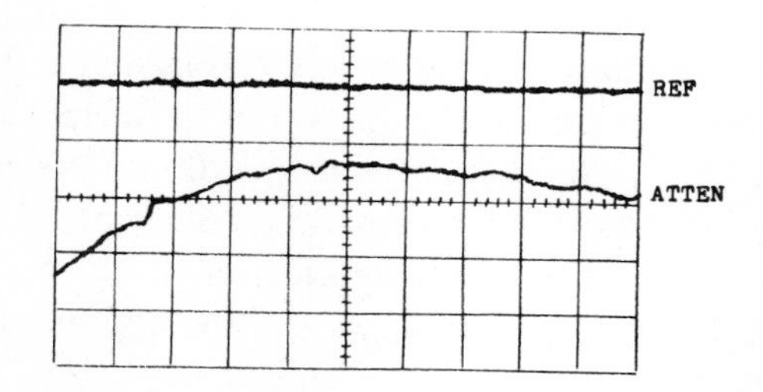

Figure 11.7 In the Swept Attenuation Method you first set up a reference (REF) level on one of the horizontal graticule lines. Then you insert the "unknown" attenuation in the circuit, and the display becomes like that at ATTEN. By increasing the vertical sensitivity of the oscilloscope you restore this trace to the original reference level. The increase of sensitivity required is the measure of the attenuation.

Before measuring the "unknown" attenuation, you set up a reference level on the oscilloscope screen, by adjusting its vertical input attenuator and the sweep oscillator power level control until the trace is positioned along one of the horizontal graticule lines. Then you insert the unknown attenuator in the circuit, which causes a decrease in the amplitude of the display. You now increase the vertical sensitivity of the oscilloscope until the trace returns to the reference position.

The increase of sensitivity required to offset the effect of the attenuator is the measure of its attenuation. This can be expressed in decibels as follows:

$$\frac{\text{Second vertical sensitivity setting}}{\text{First vertical sensitivity setting}} = \frac{E_2}{E_1} = 20 \log \frac{E_2}{E_1} \text{ dB}$$

Swept frequency measurements have the advantage that they give you the answer for the whole band that the attenuator is designed for, instead of just for a single frequency. You could, of course, make a large number of measurements at various single frequencies across the band, but this would be very tedious.

Both of the methods we've covered measure the insertion loss that resulted from connecting the attenuator into the circuit, and this is generally true whatever method we use. Impedance measurements, whether single frequency or swept, are done differently, as we shall see.

IMPEDANCE MEASUREMENTS

Impedance matching a load to its generator is one of the most important considerations in microwave transmission systems. If the load and source are mismatched, part of the power is reflected back along the transmission line. This reflection prevents maximum power transfer, can be responsible for erroneous measurements of other parameters, and may cause circuit damage in high-power applications.

Since reflections cause standing waves, the degree of mismatch can be determined by measuring the standing-wave ratio. There are two methods for doing this: the *slotted-line technique* and the *reflectometer technique*.

Slotted-Line Technique

As the name implies, a slotted line is a short section of waveguide or coaxial transmission line with a lengthwise slot. This slot is constructed to allow the insertion of a probe, while at the same time radiating little or no RF energy. Two examples are shown in Figure 11.8.

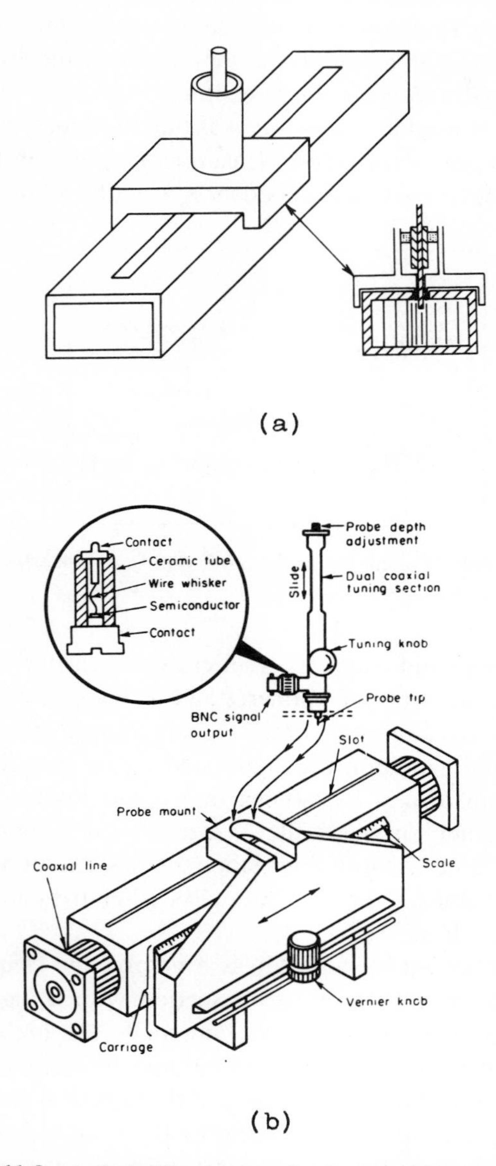

Figure 11.8 A slotted line is a length of waveguide with a slot in which a probe is inserted, as at (a). It is usually mounted in a carriage, as at (b), the probe being inserted just far enough to pick up a sample of the signal without distorting the field in the waveguide. The probe is tunable, and may have a built-in crystal detector (inset).

The slotted line is used with a carriage on which the probe is mounted in such a way that you may move it along the slot, and also adjust the depth of penetration of the probe tip (which is really a small receiving antenna). The probe should be inserted only far enough to capture the smallest usable sample of the signal in the line, otherwise you'll alter the SWR and get a false reading.

A slotted line is used in a setup such as that in Figure 11.9. It is vital, of course, as in making attenuation measurements, that all the items and their connections have the same impedance, usually 50 ohms.

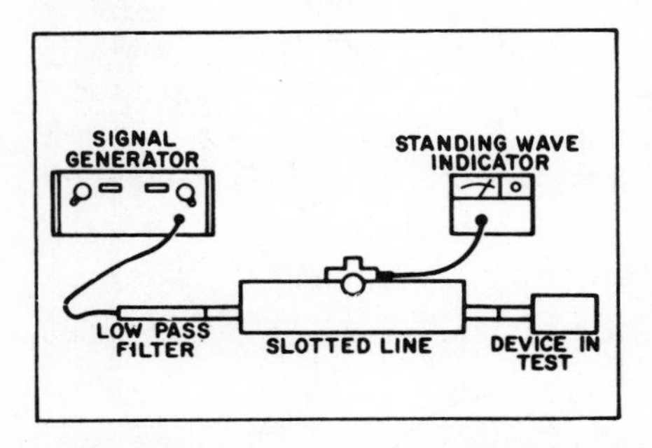

Figure 11.9 Typical Setup for SWR and Impedance Measurements in coax.

The signal generator supplies a microwave carrier at the desired frequency, modulated with a 1-kilohertz audio signal. You'll adjust this output according to the amplitude of the signal required at the slotted line.

From the signal generator the modulated signal goes through a low-pass filter to eliminate any harmonics that could impair measurement accuracy. A waveguide low-pass filter is a section of waveguide in which an arrangement of tuned obstructions stops the passage of all frequencies above the cutoff value. A coaxial low-pass filter uses tuning stubs to obtain the same result.

The next component in the setup is a microwave frequency meter. This consists of a special transmission section with a high-Q resonant cavity, tuned by a choke plunger. As you rotate the tuning knob, this plunger is screwed further out or in to enlarge or decrease the cavity. When the cavity is tuned to the frequency of the signal, it absorbs some of its power, causing a drop in the reading of the output device. The frequency is then read from the position of the index on the helical scale around the meter. The accuracy of the frequency meter is around 0.1 percent.

The attenuator that comes next needs no explanation, as it is performing the same function of isolation that we saw in the attenuation measurement setup.

The slotted-line probe contains a crystal detector to demodulate the RF signal in the same way as in the detector mount used for attenuation measurements. Its output, which is the one-kilohertz audio signal, of course, is applied to the SWR meter. The probe is moved along the slot by means of the adjusting knob until the meter gives a maximum reading. You then adjust the meter gain control until the meter pointer is positioned exactly over the 1.0 SWR mark at the right-hand end of the scale.

If you know what the SWR should be for the impedance you are checking, you'll know if it is correct because the meter reading will agree very closely with the value. However, if the impedance is really an "unknown" one, you must calculate it from the SWR, since:

$$\mathrm{SWR} = \frac{Z_L}{Z_O}$$

where $\dot{Z}_L$ is the load impedance connected at the other end of the slotted line and Z_O is the characteristic impedance of the line or waveguide setup. This is not quite as straightforward as it looks, because impedance, as we've seen earlier, is composed of both resistance and reactance, with their vectors at right-angles to each other. You can, of course, add or subtract resistances or reactances, as long as you keep them separate, which is done by writing them in the form $Z = R + jX$, where the j denotes the reactance. If this is inductive, it is positive, if capacitive, it is negative ($-jX$).

However, when it comes to dividing $\dot{Z}_O$ into Z_L, it gets a little complicated, so most people use special slide rules or charts to simplify the process. One of the most well-known is the chart illustrated in Figure 11.10, devised by Philip H. Smith of the Bell Telephone Laboratories.

To take an example, suppose your SWR reading from the meter is 2.0. You begin by drawing a circle S with center at X, and radius such that the circumference intersects the horizontal axis XY at 2.0.

The distance from the load to the first voltage minimum is read from the scale on the slotted-line carriage. Suppose this is 2.38 cm. You then move the probe along the slotted line to measure the distance between two adjacent minima.

The distance between two adjacent minima is half a wavelength. If you measured 14.88 cm., the wavelength would be 29.76 cm. (These small measurements are made easier on some slotted-line carriages by providing a micrometer screw gage.)

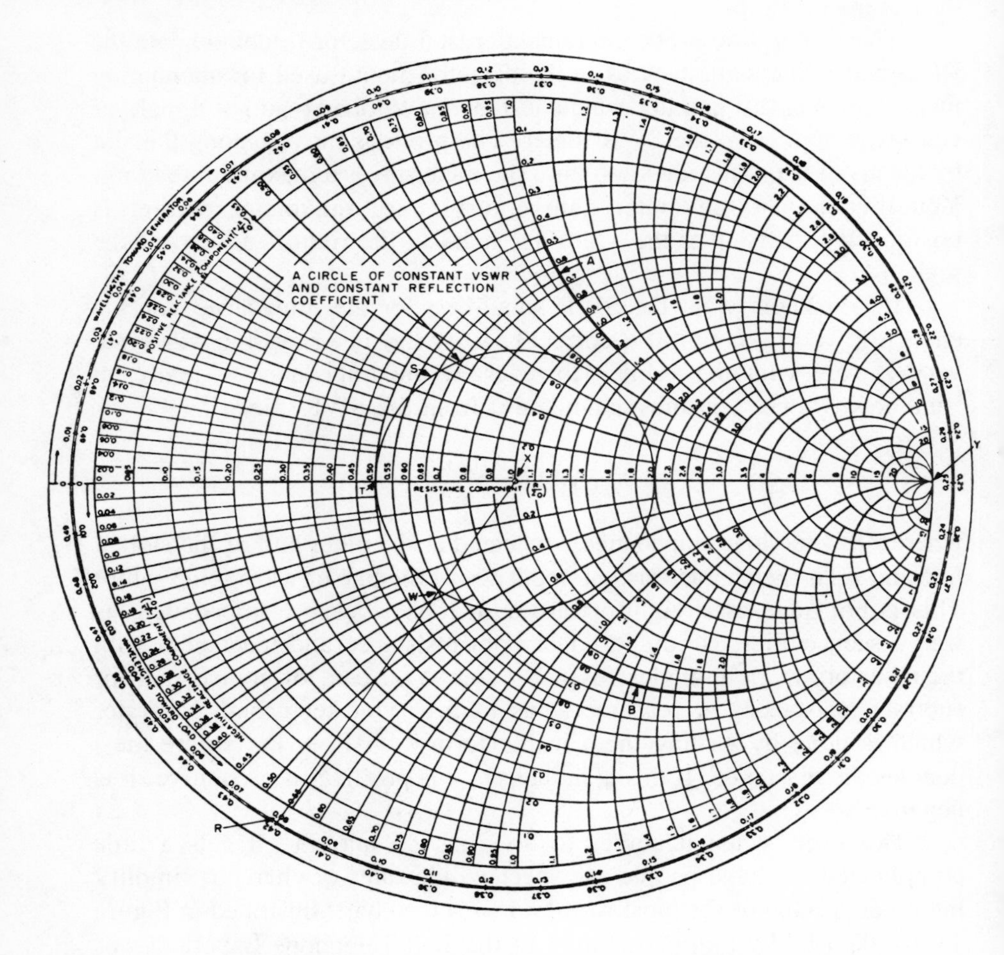

Figure 11.10 Smith Chart for Impedance Measurement

Then 2.38 cm can be expressed as 0.08 wavelength by dividing 29.76 cm into it ($\frac{2.38}{29.76} = 0.08$).

The outer circle of the chart is marked with two scales. One is wavelengths toward the generator. This one runs clockwise from zero at the left-hand end of the horizontal axis. The other is wavelengths toward the load, which, therefore, runs counterclockwise from the same point. Each scale is marked from zero to .50, since the wave pattern is spaced at half-wavelengths. In our example you'd look for 0.08 on the counter-clockwise scale, marked R in the figure.

From this point you draw a line to X. It intersects the circle S at W.

There are two "families" of circles on this chart. One family has its centers distributed along XY so that each circle is tangential to Y. The other family has its centers distributed on an imaginary line passing through Y at right angles to XY. These circles are also tangential to Y.

The first family is used for measuring the resistive component of the impedance, which is in the horizontal direction. The second measures the reactive component, which is in the vertical direction, of course. For instance, B is part of a resistance circle with the value 0.8. A is part of a reactance circle with the value 1.0. The reactive values above the horizontal axis are inductive, and therefore positive. Those below the line are capacitive, and therefore negative.

The two circles that intersect at the point W give the resistive and reactive components of the load impedance you are calculating. The resistance circle has the value 0.60, read on the horizontal axis; while the reactance circle has the value 0.38 read on the "negative reactance component" scale around the lower half of the circumference of the chart. Written in complex notation, this would be $Z_L = .60 - j.38$.

These would only be the actual values of the impedance if the slotted line's characteristic impedance was one ohm. If your slotted line has a different Z_O, the impedance values must be multiplied by it. Since this is most often 50 ohms, the true value of Z_L will probably be

$$(.60 \times 50) - j(.38 \times 50) = 30 - j19 \text{ ohms.}$$

Reflectometer Technique

The reflectometer technique for measurement of impedance is somewhat like that used for measurement of attenuation, as you can see by Figure 11.11.

The source of the test signal is a sweep generator, and the readout device is an oscilloscope. The only additional piece of equipment required is a waveguide or coaxial short. This device short-circuits the output of the second directional coupler, and because it has a reflection coefficient of one, it reflects all of the signal.

The way this setup works is to hold the forward power constant while measuring the *return loss* of the load, instead of determining the direct ratio of the incident and reflected signals. The short is used to calibrate the system by connecting it as shown in Figure 11.11. Then, with the sweep generator output at zero, you set the oscilloscope for a DC input, and adjust its vertical sensitivity for a deflection of 5 millivolts per centimeter or graticule division. With the vertical position control you then align the trace with the horizontal line 3 centimeters above the center graticule line.

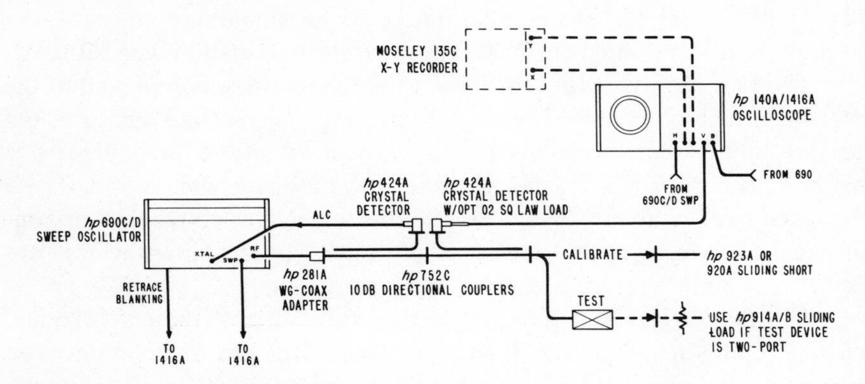

Figure 11.11 Waveguide Reflectometer with Oscilloscope Read-out *(Courtesy Hewlett-Packard Co.)*

You now increase the RF output of the sweep generator until the trace is deflected downward to 3 centimeters below the center graticule line. Don't touch the oscilloscope vertical position control or vertical sensitivity control while doing this.

The oscilloscope trace now indicates the level of the reference signal, which is 100 percent of the incident signal because it is totally reflected by the short. When you substitute the load you are testing for the calibrating short, the oscilloscope deflection will decrease. The amount by which it decreases is the return loss.

The return loss may be calculated from the formula:

$$\text{Return loss} = -20 \log_{10} \frac{\text{Test load of deflection}}{\text{Reference deflection}} \text{decibels}$$

The ratio of test load deflection to reference deflection is the ratio of the magnitude of the reflected signal to that of the incident signal, and is called the *reflection coefficient magnitude,* for which the Greek letter ρ (rho) is the symbol. The standing-wave ratio may then be calculated from another formula:

$$\text{SWR} = \frac{(1 + \rho)}{(1 - \rho)}$$

Once you know the SWR you can calculate the load impedance, as in the previous example.

However, some manufacturers of microwave equipment supply aids to help in these calculations, such as Hewlett-Packard's reflectometer slide rule, and SWR scales that you superimpose on your oscilloscope screen. Armed with these, you can practically eliminate the math.

We mentioned earlier that coaxial directional couplers don't have quite as good directivity as waveguide couplers. The type N connector used in coaxial systems increases the uncertainty in coaxial measurements, so their accuracy is less than that of waveguide systems. Recently, improved types of connector, with better impedance match, have been devised to reduce connector reflections.

POWER MEASUREMENT

As we mentioned at the beginning of this chapter, it is not easy to determine voltage or current values at microwave frequencies, since even short lengths of transmission line or waveguides are appreciable fractions of a wavelength—may easily exceed it! To make matters worse, any mismatch between generator and load results in standing waves on the line, so voltage measurements would be no better than guesswork anyway.

Fortunately, power does not vary with position, so power measurement is a very practical way to ascertain the amount of microwave energy present. Power measurement is done by determining the heating effect of the total emission, or a known fraction of it. This measurement can be of the *average power,* or the *peak power* of a periodic pulse as in radar.

There are two types of test equipment for measuring power at microwave frequencies. *Bolometric power meters* can make direct measurements of power up to ten milliwatts. *Calorimetric power meters* can make measurements from ten milliwatts to ten watts. Both types can be used to make indirect measurements of higher power.

Bolometric Power Meters

A *bolometer* is a device that measures small amounts of radiant energy. Its resistance varies with its temperature, which varies in turn with the amount of RF power absorbed. There are two types of bolometer: the *barretter* and the *thermistor*.

A *barretter* resembles a cartridge-type fuse. In place of the usual fuse wire is a sensing element consisting of a very fine platinum wire, with a diameter of from .75 to 1.5 micrometers and having a resistance of from 100 to 200 ohms. Obviously, such a fine wire has a limited current-carrying capacity, which is generally of from 5 to 10 milliamperes. The resistance of this wire *increases* with temperature, in the same way as a tungsten-filament lamp. A barretter may be used for either average or peak-power measurements, as it is a square-law device.

A *thermistor* is similar in outward appearance, but contains a small high-resistance ceramic bead (diameter about 1/100 inch) mounted on fine wires. The resistance of the bead *decreases* with temperature. It is also more rugged and can handle more power than the barretter, but it cannot be used for peak-power measurements, as it is not a square-law device.

The slope of resistance versus power is not constant in bolometers, so power measurements made by monitoring resistance are impractical. Instead, a biasing current, DC or AF, is made to flow through the thermal element to hold it to a constant resistance. When RF power is applied, the DC or AF is reduced to maintain the resistance at the same value. The amount by which it has to be decreased indicates the RF power level.

Figure 11.12 shows how a bolometer is used in a resistance bridge circuit that is self balancing. When the bridge tends to become unbalanced by the application of RF energy to the bolometer, it supplies regenerative feedback to the audio oscillator, so that less AF is fed to the bolometer, and the bridge balance is maintained. The VTVM reads the AF voltage reduction as RF power. The rheostat in the DC bias circuit is used to zero the meter, before applying RF power, by adjusting the relative proportions of AF and DC applied to the bridge.

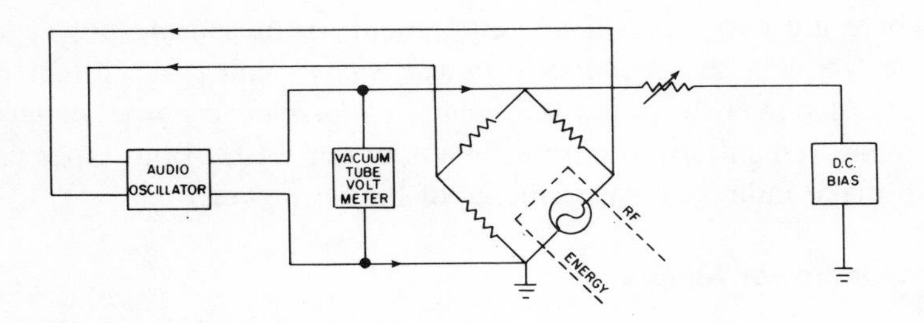

Figure 11.12 Principle of Bolometric Power Meter

The main problem with this bolometer bridge is that changes in the ambient temperature as well as in RF power can affect the bolometer, causing erroneous meter readings. To prevent this, bolometer power meters use *two* bridges. The two bolometers are positioned as close together as possible, so as to experience the same ambient temperature, but only one is in the RF field. In this way ambient temperature changes affecting both bolometers cancel each other out.

The two bolometers are installed in a *bolometer mount* connected to the main instrument by a cable. This cable conveys the AF and DC bias to the bolometers. Bolometer mounts are either coaxial or waveguide, and are connected as a termination to the transmission line, or a branch from it. To avoid power loss due to impedance mismatch, it is essential to use the right mount for the frequency. Any residual mismatch may then be tuned out with a *slide-screw tuner*. This is a slotted section of waveguide or coaxial transmission line on which is mounted an adjustable probe. The position and penetration of the probe is adjusted to set up a reflection that cancels out existing reflections.

A bolometric power meter can only measure power absorbed by the bolometer, not that portion dissipated in the amount or reflected by it. Therefore, there is a difference between the total RF power and the value shown on the meter. (The meter reading, of course, is really the value of the substituted bias power.) The ratio between the two values is called the *calibration factor*. Some power meters have a calibration factor control that you set to the calibration factor of the bolometer mount in order to get a corrected meter reading.

The ratio of substituted bias power to microwave power absorbed by the mount is called *effective efficiency*. Both factors are required to obtain accurate readings at low power levels.

Calorimetric Power Meter

Calorimetric power meters may be *flow* or *static* (liquid or dry). The principle of operation is the same in each—to dissipate the microwave energy as heat and then measure the resulting temperature rise. Test-equipment models are usually liquid. Figure 11.13 illustrates a flow calorimeter in which the liquid is oil.

The oil flows from the pump through the heat exchanger to the *comparison head*, back to the heat exchanger, then to the *input head*, and finally back to the pump. The reason for the two trips through the heat exchanger is to make sure the oil is at the same temperature as it enters each head. An equal rate of flow is also obtained because the pump drives the same oil through each head in turn.

Both heads contain a load resistor and a temperature-sensitive resistor. These are immersed in the oil stream in such a way that the oil flows around the load resistor before it reaches the temperature-sensitive resistor.

The input RF heats the load resistor in the input head in proportion to the power of the signal. This heat is carried by the oil to the input temperature-sensitive resistor, which therefore increases in temperature with a corresponding alteration in its resistance value. As you can see from Figure 11.13, this resistor forms one arm of a bridge. The comparison temperature-sensitive resistor forms another. As long as both resistors are at the same temperature their resistances are equal, and equal AF currents flow through them. But when the input temperature rises, the bridge becomes unbalanced, and the currents are no longer equal.

This results in an unbalance signal at the input of the amplifier, which develops an output current in porportion. This current heats the comparison load resistor to exactly the same temperature as the input load resistor, so that its heat, carried by the oil, causes the comparison temperature-sensitive resistor to change its resistance to the same value as the input resistor, so that the bridge is rebalanced.

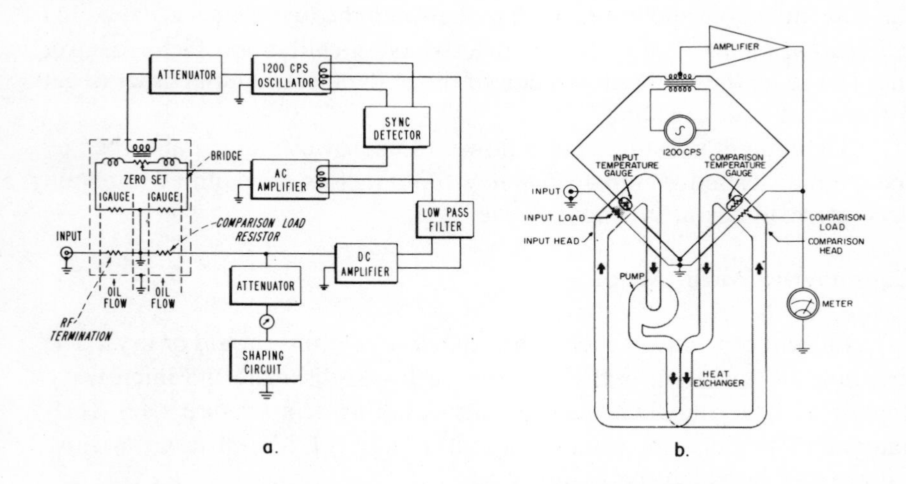

Figure 11.13 Block diagram (a) and oil flow diagram (b) of hp 434A Calorimetric Power Meter. Moving oil stream and bridge-feedback arrangement results in fast response time to input power changes enabling dynamic tuning of units in test. Series oil flow arrangement eliminates flow rate errors usually encountered in flow calorimeters. *(Courtesy Hewlett-Packard Co.)*

The output current from the amplifier required to rebalance the bridge is measured by the meter. Since this current is directly proportional to the RF input power, the deflection of the meter pointer along a scale calibrated in watts will give the microwave power of the signal. This calorimeter power meter can measure up to 10 watts of average power, or a kilowatt of peak power (as in a radar signal). It has an attenuator to select full-scale meter ranges from 10 milliwatts to 10 watts. You can measure higher power by using an external attenuator or directional coupler, of course, in which case you would multiply the reading by the attenuation factor.

There are two calibration adjustments you must set before using this meter. The *zero set* adjustment balances the bridge by equalizing the bias currents in the two arms. The meter will read zero when you have set this correctly.

The other adjustment is made with an internal DC signal of 100 milliwatts applied. This time you adjust the meter pointer for a full-scale reading, with the range control set to the proper range. When this precalibration has been done, you can use the meter to measure the power of microwave signals up to 12.4 gigahertz with an accuracy within five percent.

12

TROUBLESHOOTING TV WITH TEST EQUIPMENT

Every defect in a TV receiver has to affect the picture or the sound, or there wouldn't *be* a defect. The picture tube and the loudspeaker show us the end product of all the circuits in the set. The way in which this differs from the normal indicates the section of the set where the trouble is most likely to be. Therefore, the TV set is its own analyst, continuously testing itself, and is the first, and sometimes the only, piece of test equipment you need.

However, in order to isolate the defective component the television service shop should have the following items as a minimum requirement:

- Color bar generator
- VTVM or VOM
- Oscilloscope
- Sweep generator
- Tube/transistor tester
- Impedance bridge (capacitance/inductance tester)
- Schematics and alignment instructions for each set serviced

Other equipment, such as a CRT tester or a field strength meter, are not absolutely necessary, but can be very helpful and timesaving.

We have already discussed most of these in earlier chapters, so this chapter will concentrate on the TV set as a test instrument, and the use of the color bar generator and the schematic diagram.

THE TV SET AS A TEST INSTRUMENT

In Figure 12.1, we show the loudspeaker and picture screen of your color TV set. The loudspeaker has only one function, of course, the conversion of audio signals into sound waves, but the picture tube has three different ones: the generation and deflection of the electron beams that activate the phosphor screen, and the modulation of these beams with picture information in both "whiteness" and color. In a monochrome set, the third function is absent, of course.

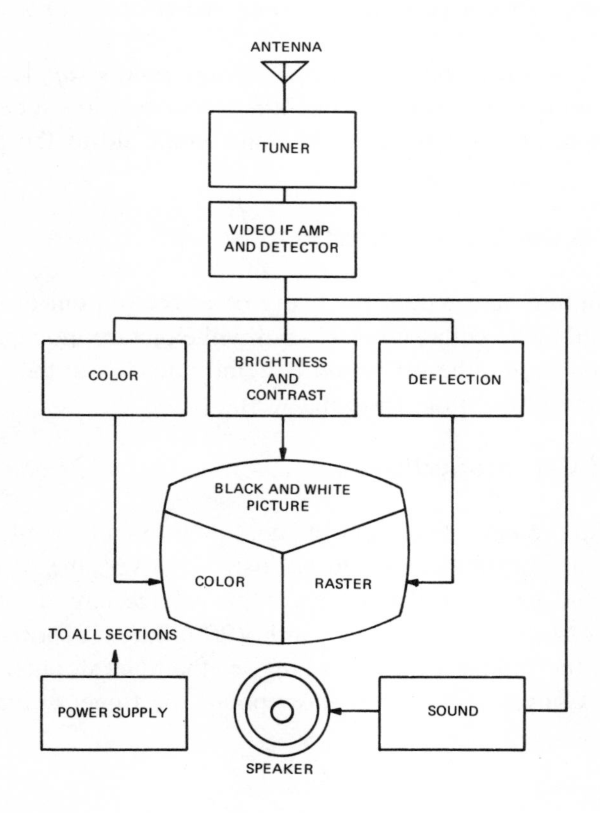

Figure 12.1 Principal Sections of a TV Set and Their Outputs

Each of these functions requires a major section of the set to develop the necessary driving signals or voltages. The *sound section* amplifies and demodulates the FM audio signal, and gives it sufficient power to drive the loudspeaker. The *raster section* generates the vertical and horizontal deflection signals that are applied to the vertical and horizontal yoke coils, and also the accelerating voltages that give the electron beams the energy to excite the phosphor screen. This section is also responsible for synchronizing its deflection signals with those of the incoming TV signal, so that an intelligible picture results. The *black-and-white section* handles the brightness (or Y) signal of a color broadcast (in a monochrome receiver the black-and-white video signal), and the *color section* processes the color information. The TV signal captured by the antenna has also to be tuned-in, amplified and demodulated before being distributed to the various sections. This is done in the *tuner* and *IF/detector sections,* of course.

There is one other section, *the low-voltage power supply,* that provides the operating voltages for all the other sections. Now let's see what a total failure of any one of these sections would do to the picture or sound.

Low-Voltage Power Supply (LVPS)

A total failure here would put every other section out of action, so the effect would be no color, no black-and-white picture, no raster and no sound. In other words, the set would be totally dead, just the same as if you'd pulled the power plug from the wall.

Tuner, IF and Detector Sections

A total failure here would cut off both picture and sound, since the video and audio information would be lost. However, the raster would still be produced by the deflection and high-voltage circuits, so the set would not be completely dead, as in an LVPS failure. If snow is present on the screen, the IF amplifier is working, so the tuner or antenna would be suspected. Conversely, lack of snow points the finger at the IF amplifier.

Audio Section

Total failure here would result in no sound, but the picture would be normal.

Raster Section

A total failure in this section would black out the picture screen, but would not affect the sound.

Black-and-White Section

In a monochrome set, this section comprises the video amplifier, contrast and brightness controls and DC restorer. Total failure here will wipe out the picture, but leave the raster and sound unaffected. This situation resembles that where the failure is in the IF amplifier. However, if the latter were at fault, the raster would be unsynchronized, because the sync pulses would be lost along with the picture information.

How can you tell if the raster is unsynchronized? Well, as you know, a normal raster consists of 525 horizontal lines. These really consist of two fields of 262½ lines each, with the lines of the second field interlaced between those of the first. However, if the sync pulses are missing, they will not be interlaced properly, so the presence of this symptom will suggest to you that the problem is a tuner or IF failure rather than a video one. To confirm this, you can adjust the vertical hold control to place the vertical blanking bar in the middle of the screen, and adjust the brightness control so you can see if sync information is present on the blanking bar. This is in the form of about six darker lines with a dark area in the center called a "hammer head." If these are present, your set is processing the video signal up to the takeoff point for the sync information, thus confirming that your trouble is between that point and the picture tube.

In a color set, it is possible to lose the black-and-white picture information but still have a color picture, although the colors will be vague and smeary. This is because the defect lies in the Y amplifier, after the point where the color signal is taken off. The reproduction of color is not impaired, but the hues lack outline and brightness because the Y signal is absent.

Color Section

A total loss of color in the picture when a color program is being received must mean a complete failure in the color section alone, if a satisfactory picture is displayed on a monochrome program.

Table 12-I summarizes what we've just been discussing. The sections listed are composed of many circuits, of course, and many defects other than total failure are possible. However, the nature of the defect will generally point to the section and, in many instances, the circuit involved. In some cases, the same symptom could be caused by a defect in more than one section. For example, the condition of no raster could be due to trouble in the raster section or in the LVPS. But in the latter case, the sound would be absent also.

TABLE 12-I

MAJOR SYMPTOMS	Section Most Likely to be in Trouble							
	LVPS	Raster	Video (Y) Amp	Color	Sound	Tuner	IF	Picture Tube
Dead Set	?							
Screen dark, sound OK		?						?
No picture or sound; raster OK			?				?	
No picture or sound; snow						?		
Picture OK, no sound					?			
No color; black & white, sound OK				?				
Color's vague & smeary, sound OK			?					

So far we've considered only the case of a totally inoperative low-voltage power supply, which results in a dead set. This is the easiest condition to deal with, because there has to be some complete breakdown between the power line and the power supply output. Figure 12.2 shows the principal elements in a typical power supply. With a voltmeter it is a fairly simple task to determine where the interruption, is and then to identify the cause.

Table 12-II lists several other common symptoms where the set is not dead but is obviously sick. The first four indicate a low B+ or V_{cc} voltage, so you'd begin by verifying that the cause was really in the power supply and not in some other section, where a lowered resistance to ground was loading down the supply.

As we shall be going into troubleshooting in more detail in Chapter 15, we'll not go further into tracking down power-supply problems here. The purpose of this section is to demonstrate the use of the TV set in diagnosing its own troubles.

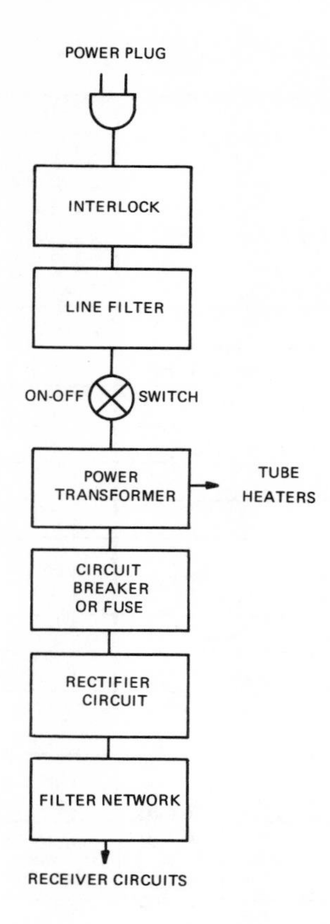

Figure 12.2 Principal Elements in a Typical Power Supply

Sound Section

Next to the LVPS, the sound section is the easiest to analyze, especially if you are familiar with FM receivers. All modern TV sets have 4.5 megahertz intercarrier sound. This is separated from the picture and sync information at a point between the last video IF stage and the video amplifier. In color sets you will often find a sound detector provided in addition to the picture detector. This does not recover the audio signal, which is done in the FM detector in the sound section, but serves to keep the sound signal out of the video amplifier.

The stages of the sound section are shown in Figure 12.3, and some common symptoms and their probable causes are given in Table 12.III.

TABLE 12-II

LVPS SYMPTOMS	Circuit Most Likely to be in Trouble							
	No line voltage	Low line voltage	On/off switch	Power transformer	Circuit breaker or fuse	Rectifier circuit	Filter network	Receiver circuits
Dead set	?		?	?	?	?	?	
Dark, no raster, some sound		?				?	?	?
Blooming		?				?	?	
Brightness lacking		?				?	?	
Size lacking		?				?	?	
Hum bars				?		?	?	
Wavy, vertical				?		?	?	
Vertical sync critical				?		?	?	
Intermittent operation			?	?	?			
Hum (not from speaker)				?				

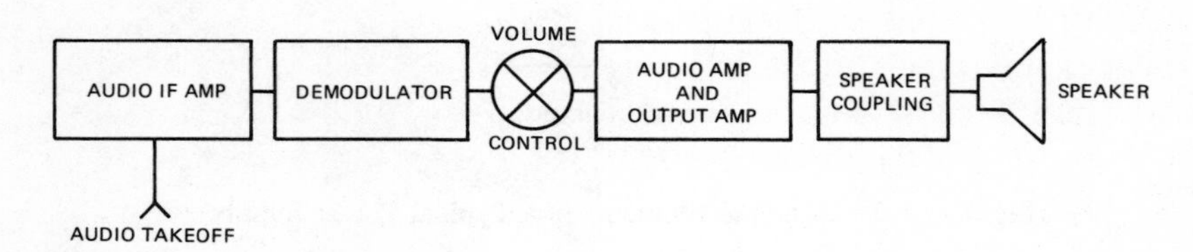

Figure 12.3 Principal Elements in a Typical Sound Section

Raster Section

This is a large section that is subdivided as shown in Figure 12.4. The subdivisions make up two main groups, the sync and deflection sections. Table 12.IV lists the most common symptoms of trouble, and their most likely causes.

Sweep and High Voltage Power Supply defects may be due to the LVPS, the HVPS, or the deflection circuits. If the vertical or horizontal size of the raster is reduced, the most likely culprit is the LVPS, but if the LVPS employs solid-state diodes, it may save you work if you try new vertical or horizontal output tubes first. The damper and horizontal oscil-

TABLE 12-III

AUDIO SYMPTOMS (Picture OK)	Circuit Most Likely to be in Trouble							
	Sound takeoff	Audio IF Amp	Demodulator	Volume Control	Audio Amp	Audio Output	Speaker coupling	Speaker
Sound absent or weak	?	?	?	?	?	?	?	?
Intermittent sound	?	?	?	?	?	?	?	
Hum		?	?		?	?		
Buzz	?		?					
Motorboating ("putt-putting")					?	?		
Hiss			?		?	?		
Rasping or crackling		?	?	?	?	?		?
Ringing, whistling, howling, &c					?	?		
Distortion	?		?			?		?

lator tubes also should be checked, since they can affect width as well. If none of these is bad, or if the set is a solid-state one, you would then measure the supply voltages, as explained above, and also the boost voltage. However, if these turn out to be normal, the cause *must* be in the deflection circuits, so you would examine the waveforms in these circuits, especially the drive signals for the output stages.

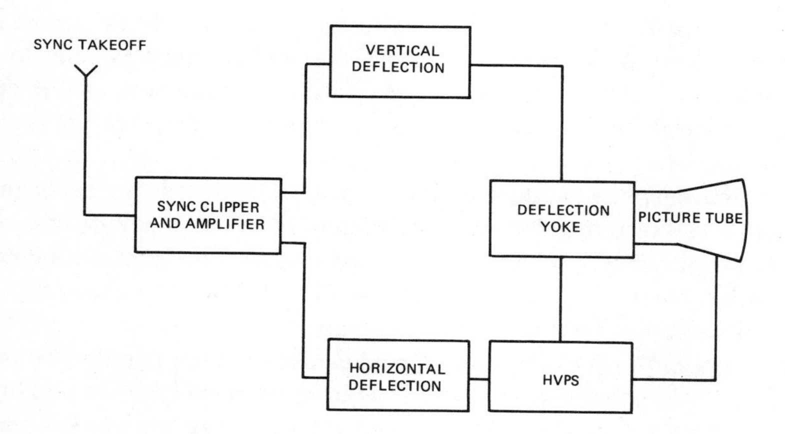

Figure 12.4 Principal Elements in a Typical Raster Section

TABLE 12-IV

RASTER SYMPTOMS	Circuit Most Likely to be in Trouble							
	Sync clipper/amp	Vertical defl'n.	Horiz. defl'n.	HVPS	Deflection yoke	Picture tube		
Sync critical or lacking	?							
Compressed, top or bottom		?		?				
Vertical fold, excessive height		?						
Insufficient height		?			?			
Rolling vertically		?						
Single horizontal line		?			?			
Dark, no raster			?	?	?	?		
Brightness excessive			?	?				
Slanting lines, unsynchronized			?					
Jittery or jumping			?	?				
Width excessive or insufficient			?	?	?			
Blooming, poor focus				?	?	?		
Brightness lacking				?	?	?		
Wedge-shaped					?			
Bending at top	?							
Pulling	?							

Non-linearity, because it is caused by variations in the speed at which the electron beams travel across the screen, must be due to a deflection circuit defect. In most cases, careful readjustment of the appropriate controls will correct it. On the other hand, *pincushioning (concave) or barreling* (convex), where the electron beams fail to follow a straight line near the borders of the screen, are caused by the beams having to travel different distances to different parts of the picture area. In black-and-white receivers, this is corrected by small magnets on the rim of the yoke. Displacing these may cause this condition. In color sets, a correction circuit is used instead of magnets.

A *trapezoid* raster, in which opposite edges are not parallel, is usually caused by a shorted yoke coil. A picture that gets bigger as you turn up the brightness control is said to *bloom*. It also gets less bright and goes out of focus. The cause is poor regulation in the HVPS, and you should check the HV rectifier tube for low emission. The trouble, however, is sometimes due to a weak damper tube or horizontal output tube, or even to low B+ voltage.

Another raster trouble sign is *foldover*. This may be vertical or horizontal. The bottom or sides of the picture are "folded" back over the edge of the main picture, giving a brightened margin in which the image is inverted or reversed. Where the foldover is at the bottom, suspect first a gassy vertical output tube, or leakage in the capacitor coupling it to the previous stage. Either defect will produce a non-linear sweep. If the foldover is in the right half of the screen, you probably have similar troubles in the horizontal output stage, but if the left half of the screen is involved, you should check the damper tube or yoke circuit.

Another type of distortion is *bending,* where the raster is pulled sideways, so that the vertical edge is wavy (like an S), or hooked. This condition may also be caused by a defect in the sync, video or IF circuits, so you should check by removing the video signal. If the plain raster still shows the same symptoms, you probably have poor filtering in the LVPS or boosted B+, or a heater-cathode leak in a horizontal deflection-circuit tube.

Other raster troubles are caused by sync problems. These range from slight unsteadiness to complete loss of control. The condition may affect vertical or horizontal deflection, or both. Loss of *vertical* sync results in a picture that rolls vertically. If this cannot be corrected with the vertical hold control, the trouble may be loss of the sync signal or a defect in the vertical multivibrator. If you can get the picture to roll slowly enough so that you can check the blanking bar, as explained above, you can see if sync pulses are getting as far as the sync takeoff point. This test also will demonstrate that the vertical multivibrator can free-run at approximately the proper rate. The trouble is, therefore, probably between the sync takeoff point and the vertical deflection circuit. This section (see Figure 12.4) processes the sync pulses for both vertical and horizontal sections, but if the loss of sync is only partial, the vertical may be affected while the horizontal continues to hold. Of course, if *both* vertical and horizontal hold are lost, the evidence is even more definite. When horizontal sync is lost, and adjusting the horizontal hold control will not restore it, the problem is usually in the horizontal oscillator or AFC circuits. If vertical sync is lost also, as mentioned above, the trouble must be in the sync circuits common to both.

In some cases, the picture remains synchronized but is marred by jitter or distortion. The cause of this most likely is noise or interference, due to high-voltage arcing or corona discharge, a defective filtering circuit somewhere in the sync path or in the LVPS, heater-to-cathode leakage in a tube, or video information in the sync signal. The last may be due to trouble in the IF, video amplifier or sync separator.

Video Defects point to trouble in the signal path from the antenna to the picture tube (Figure 12.5). We've already mentioned a completely blank raster, indicating a complete break in this signal path, and that the presence or absence of sync tells you whether this is before or after the sync takeoff point. Another clue can be gotten from the brightness control. If operating this produces no change in brightness, you may have a bad picture tube. Check the contrast control also: sometimes Junior has been playing with them!

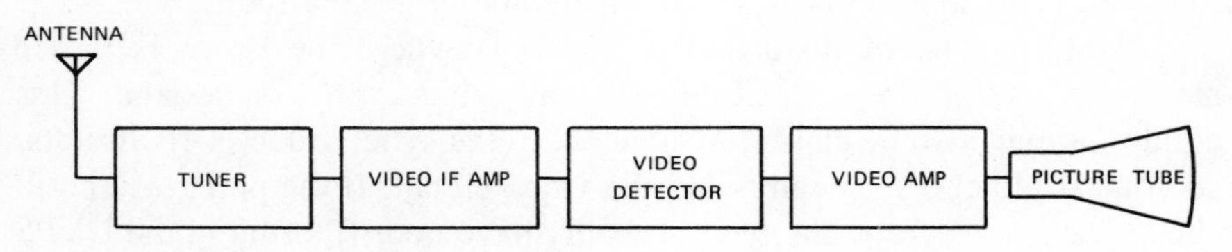

Figure 12.5 Principal Elements of a Typical Video Signal Handling Section

TABLE 12-V

SIGNAL CIRCUIT SYMPTOMS	Circuit Most Likely to be in Trouble							
	Antenna	Tuner	Video IF Amp	Video Detector	Video Amp	Picture Tube		
Snow	?	?						
Weak picture			?	?	?	?		
Lack of fine detail					?			
Smear					?			
Ringing					?			
Blank raster			?	?	?	?		

Snow on the screen, with or without a weak picture, usually means the IF amplifier is working, but there is an inadequate signal input to it. Your trouble may be a disconnected antenna or a defective tuner.

However, a weak picture *without* snow generally exonerates the tuner, and suggests the trouble is in the IF amplifier, detector or video amplifier. Again, check the contrast and brightness controls, and also the AGC control if there is one. Trouble in the video amplifier is usually evidenced by lack of fine detail, smear, or ringing (vertical edges of objects in the picture followed by several weaker edges).

Some picture defects caused by interference originate outside the set, some within. You can determine which it is by disconnecting the antenna lead-in and grounding the antenna terminals. If the interference is still there, it must be originating in the receiver. A common cause is a corona discharge. Otherwise, the nature of the interference pattern may point to the external cause (for example, automobile ignition, electrical equipment, FM broadcast, adjacent channel interference, ham radio operating, and so on).

Color defects may also indicate the source of the trouble, if carefully studied. Figure 12.6 shows the principal circuits in the color section.

No color or weak color can result from a weak signal reaching the

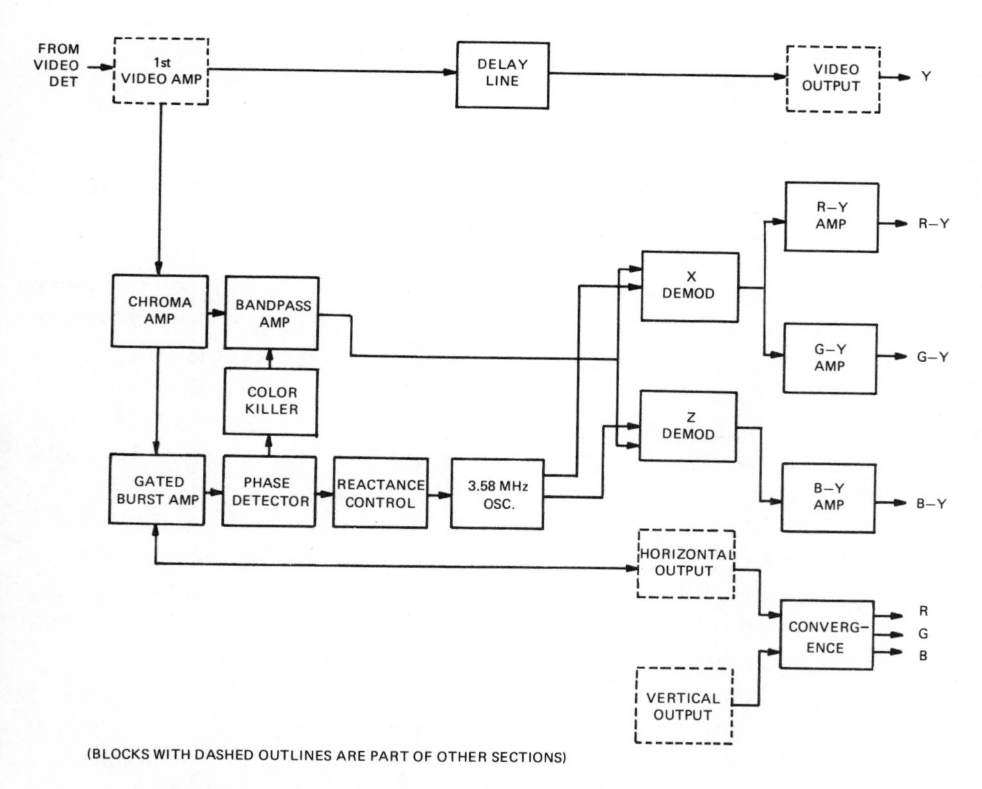

Figure 12.6 Typical Color Section (Blocks with dashed outlines are part of other sections.)

color section, but if there is good fine detail on the screen, you know the signal is adequate, and therefore the problem must be in the color section. To verify this, tune to an inactive channel and disable the color killer by rotating the color control fully clockwise. If you get colored snow, the color section is working; if the snow is white, the color section is not working.

TABLE 12-VI

COLOR SYMPTOMS	Circuit Most Likely to be in Trouble								
	Color Diff. Amp.	Color Demod.	Color Killer	Chroma or Band-pass Amp.	3.58MHz Osc.	Delay Line	Burst Amp. or Phase Detector	Reactance Control	Convergence
One color missing	?								
Colors in wring proportions		?							
Color cast on black & white pix	?								
Colors correct but weak				?					
No color at all			?	?					
Strong colors of wrong hue					?	?			
Color sync bad						?	?	?	
Multicolored dots (dot pattern)									?

It is easier to analyze color defects if you are working with a known signal. You then know what you should be getting, so any departure from normal is a clue. The color bar generator gives you such a signal, as shown in Figure 12.7. These are the RCA simulated ten bars, as generated by all modern color bar generators. Such a generator will also produce other patterns, but the color bars are the most important for troubleshooting the color section.

As you can see, the color bar generator causes the TV set's screen to display (when it is working properly) a bar of color for every 30 degrees of phase shift in the color signal. It does this by generating two signals, which are out of phase by the proper angles for each color, and gating them so that they are separated into bars of different colors.

When one of the three primary colors is missing, its absence is obvious because this color is not present in those color bars where it should be. If red is not there, the bars will have only the green and blue elements; if blue is missing, only the red and green will appear; and if green is missing, the surviving colors will be red and blue.

The loss of only one color tells you there is most likely a defect in the corresponding color difference amplifier. (However, if there should be *three* color demodulators, which is not usually the case, it could also be one of these. It could be a picture-tube circuit defect as well.)

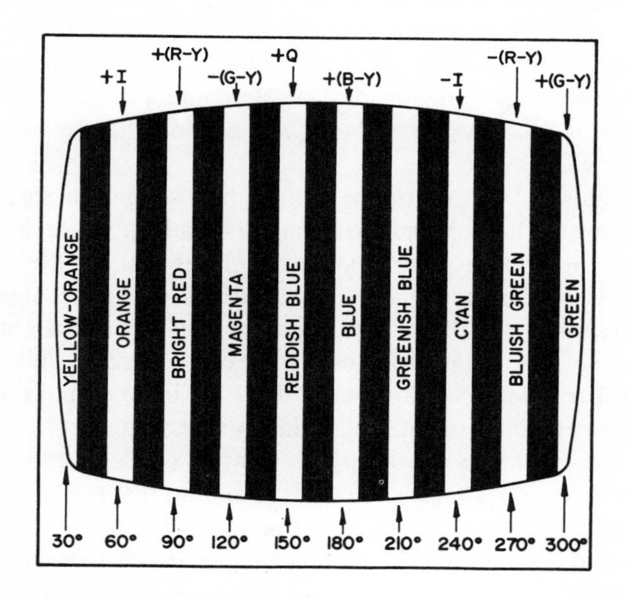

Figure 12.7 Color bar patterns showing colors, phase relation, and associated color signals

The color bar pattern may also reveal a condition where all colors are present, but not in the proper proportions. This would be the case where there are the usual two demodulators, X and Z, and one is not working. If the red and green are weak, so that blue predominates, the X demodulator is responsible. If the picture is a grayish-orange, the blue and green are weak, pointing to the Z demodulator as the culprit.

When the defective demodulator is still working, though inadequately, it is harder to recognize the cause. It may seem that the hue control needs adjusting. You'll find, however, that as you correct one color bar the others get worse. The best way to determine the cause is to check the color difference signals at the picture tube grids with an oscilloscope, comparing their appearance on the CRT screen with the way the manufacturer's schematic shows they're supposed to look.

Another indication of an upset color balance is the effect it has on a black-and-white picture. If the picture has a color tint or cast, it will be because the complementary color is weak or absent. For example, an overall purple cast, due to a predominance of red and blue, indicates the green is weak or absent, so you'd suspect the G-Y amplifier, assuming you'd first checked the color controls and found them operating normally.

If the colors are weak, although correct, and turning up the color control does not give you saturated colors, the signal cannot be getting enough amplification. Since all three colors are affected equally, the trouble must be somewhere ahead of the demodulators. The killer circuit must be working, since you are getting some color, therefore the color IF is the most logical suspect.

Good saturated colors of the wrong hue suggest a serious phase error, if the correct hues cannot be obtained by adjusting the hue control. All the colors are present, but in the wrong places in the color bar pattern. The prime suspect here is the 3.58-megahertz reference signal at the demodulator, since it is the phase difference between this signal and the color signal that determines the hue. However, a shorted delay line will also cause misregistered colors, but in this case there would probably be an additional complaint of multiple images in the picture.

When the color bars break up into slanting bands of color, that may also drift, with an appearance somewhat similar to that produced by a loss of horizontal sync, you have a color-sync problem. The 3.58-megahertz oscillator signal is not locked in with the color burst. The trouble could be in the oscillator, the reactance control or phase detector (in APC systems), or the burst gating amplifier. It will help you to analyze the problem if you ground the grid of the reactance control tube or equivalent point in a solid-state circuit, when the trouble appears in a set employing an APC system. If you find that you can get a correct color bar pattern, even if only momentarily, by tweaking the reactance coil tuning adjustment, then the reactance and oscillator circuits are functioning properly, so your trouble must be in the phase detector or burst amplifier.

Other Uses of Color Bar Generator

In addition to color bars, the color bar generator can give you white dots, horizontal lines, vertical lines, cross-hatch (horizontal and vertical lines combined) and, in some cases, single dots and lines.

Of these, the dot pattern is the most important. A standard dot pattern consists of ten rows of ten dots each. The dots should be small (one line width) and circular. They are used to perform dynamic convergence after the purity and static convergence adjustments have been made.

When the electron beams are properly converged, the dots will be white, or of a uniform color. The beams are not in register if the dots are multicolored. It is also likely that the degree of misconvergence varies in different areas of the screen.

Corrections are made by adjusting the controls on the dynamic convergence board of the TV set. Full instructions for every set are given either on the board or in the manufacturer's service instructions. These instructions may call for you to use a cross-hatch pattern instead of a dot pattern, or both alternately. Generally, it is simpler to disable the blue electron gun while converging the red and green (to give uniform yellow dots), then to re-energize the blue gun and use the blue controls to converge the blue with the yellow.

We've covered a lot of ground in this discussion of how the TV set is its own best diagnostician in pinpointing the location of the trouble, assisted by the color bar generator for color problems. However, do not lose sight of the basic procedure which you should use on every TV set, regardless of make or model:

A. Listen to the owner's description of the problem, and let him demonstrate it by operating the set. Ask questions, if necessary, such as:

(1) When did you notice it first?

(2) Does it do it all the time?

(3) Did it get progressively worse, or did it come on suddenly?

(4) Are there other troubles, apart from this one?

B. Tune in each locally active channel, and check performance. Is only one channel giving trouble, or are they all? Adjust the fine tuning and picture controls for best results. If one control makes it better (or worse) you have a clue to the circuit at fault (or maybe it's the control itself).

C. Check what happens when you operate any service adjustment affecting the section that seems to be in trouble. For instance, if the picture lacks horizontal sync the horizontal hold and frequency adjustments should be tried.

D. If the set uses vacuum tubes, always try replacing those in the affected section before checking other components.

E. Check the antenna and its connections. It's amazing how many people will buy an expensive color TV set and a brand new color antenna, and connect them together with cheap unshielded twin lead draped over metal gutters, taped to water pipes and what-have-you, and wonder why their pictures have colored fringes, snow, poor definition, washed-out colors, and so on.

Other Equipment

While the television receiver can tell us the section where the trouble is, and even the circuit, we shall probably have to use other equipment to determine which component is at fault. The most important items were listed at the beginning of the chapter. There is one further item that is extremely important, the service literature on the receiver.

Service Literature

Most manufacturers publish a service manual for their sets. This will contain the schematic diagram for the receiver, a pictorial layout giving the location of the tubes, controls, service adjustments and other features, and detailed instructions for adjusting the controls and setting up the receiver. If you can't get the manufacturer's service literature, you can get a schematic diagram and servicing information for the set from a diagram service such as Howard W. Sam's "Photofacts."

This service information usually contains detailed alignment instructions.

In general, there are two methods of alignment in common use. They are called *peak alignment* and *sweep alignment*. The first requires a signal generator and a VTVM. Each of the tuned circuits is then peaked to a specified frequency, to which the generator is tuned. The VTVM is connected to read the demodulated output at the video detector or other point given in the alignment instructions. This method is somewhat approximate, and is often done as a preliminary to sweep alignment rather than by itself.

Sweep alignment allows you to see the response of the amplifier on the oscilloscope screen, when connected to a sweep generator, as shown in Figure 12.8. The sweep generator is set to give a suitable sweep that encompasses the frequency band of the amplifier, and you compare the oscilloscope display to the response curve down in the alignment instructions. This illustration also shows what markers to use to identify the various points on the response curve. The instructions give precise directions that must be followed exactly if you want to do it right, but assuming you do so you can make the amplifier's response curve look as much

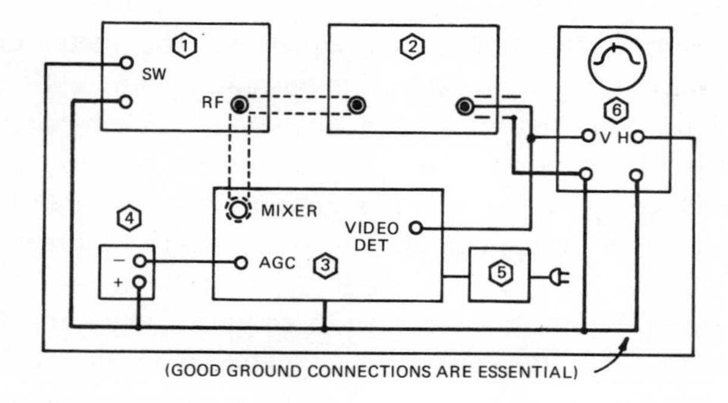

(1) Sweep Generator: using 75-ohm coaxial leads, connect RF output to ungrounded tube shield over mixer tube in tuner, or to test point provided in transistorized tuner: and also to marker generator input. Connect test leads from sweep output (60Hz) to oscilloscope horizontal input (H). Adjust sweep generator frequency to 44 MHz (10 MHz sweep).

(2) Marker Generator: connect output to oscilloscope vertical input (V). This is called "post injection," and avoids having the markers go through the circuits being aligned.

(3) Television Receiver: connect bias box to AGC bus, and adjust for bias voltage specified. Connect lead from video detector output to vertical input (V) of oscilloscope. Connect to power line using isolation transformer. Allow 20 minutes warm-up.

(4) Bias Box: none-volt battery with potentiometer.

(5) Isolation Transformer: 1:1 power transformer for safety.

(6) Oscilloscope: Adjust control for stable display similar to that shown. If signal is noisy connect a 47-ohm resistor to series with vertical input and lead from video detector, with .05-microfarad capacitor from vertical input to ground.

Figure 12.8 How to Connect Your Test Equipment for TV Alignment

like that in the alignment instructions as possible by adjusting the tuned circuits as specified. It cannot be too strongly emphasized that (a) you must follow the directions implicitly, and (b) you must use the proper equipment, which means a suitable sweep generator, oscilloscope and marker generator.

The same procedure is used for tuner, IF amplifier and sound IF. The details will vary, of course, according to the frequencies and bandwidth.

CRT Tester

The cathode-ray tube tester is a specialized type of tube tester. Black-and-white picture tubes, which have only one electron gun, of course, are tested by connecting the second grid (G2) as the plate of a tube, and measuring emission (see Chapter 5). The meter then reads BAD-?-GOOD according to the emission current. Some CRT testers also have a circuit for rejuvenating a cathode with weak emission. This is accomplished by discharging a large capacitor connected in series with the CRT impedance, the value of which gives a time constant that permits the proper voltage to be applied for the right length of time to restore adequate emission.

In the case of color tubes we have three guns, which must not only have good emission, but must also "track." This means that their cathode currents must not vary more than one third from each other at any setting of the brightness control. If they do not do this, the background color will change with different settings of the control. A CRT tester for color tubes should, therefore, provide a means of testing for correct tracking and indicating bad tracking on the meter if the ratio of any one gun to the others exceeds 1.5 to 1. A color tube tester will also test monochrome tubes.

13

HOW TO TEST AUDIO AND HI-FI EQUIPMENT

Tastes differ so much that what pleases one listener may be quite distressing to another. A booming bass sounds good if you got your musical education from a juke box, but not so good if you graduated from a musical academy. Fortunately, you do not have to be a maestro to service hi-fi equipment. Test records, a few test instruments and a critical ear are all you need. The test instruments generally used are:

- Oscilloscope
- AC VTVM
- VU Meter
- Distortion Analyzer

Test records are phonograph records that serve as a source for signals to test frequency response, distortion, noise level, wow, rumble, stylus wear, equalization, sounds of various musical instruments and so on. They can also be used to demonstrate system performance to the owner of the equipment, because their standard signals get around the difficulty where his concept of audio quality differs from yours.

Frequency response is the measure of the effectiveness with which the amplifier or other device responds to the different frequencies applied to it. A test record has a series of bands generating tone signals in individual steps between 30 and 12,000 hertz. Simply by listening to the loudspeaker you can judge the approximate response of the amplifier more accurately than by using a meter to measure the output, because the human ear has a very non-linear response to frequency, and it is to the ear that the output has to sound right.

Frequency response can also be measured with a recording that sweeps the range between 50 and 10,000 hertz. The amplifier output is viewed with an oscilloscope on which the frequency response curve appears with markers to permit frequency identification. This is similar to TV IF amplifier alignment.

Distortion is any undesirable change in the waveform between the input and the output of the amplifier. Three fundamental types of distortion are amplitude, frequency and phase distortion (see Figure 13.1).

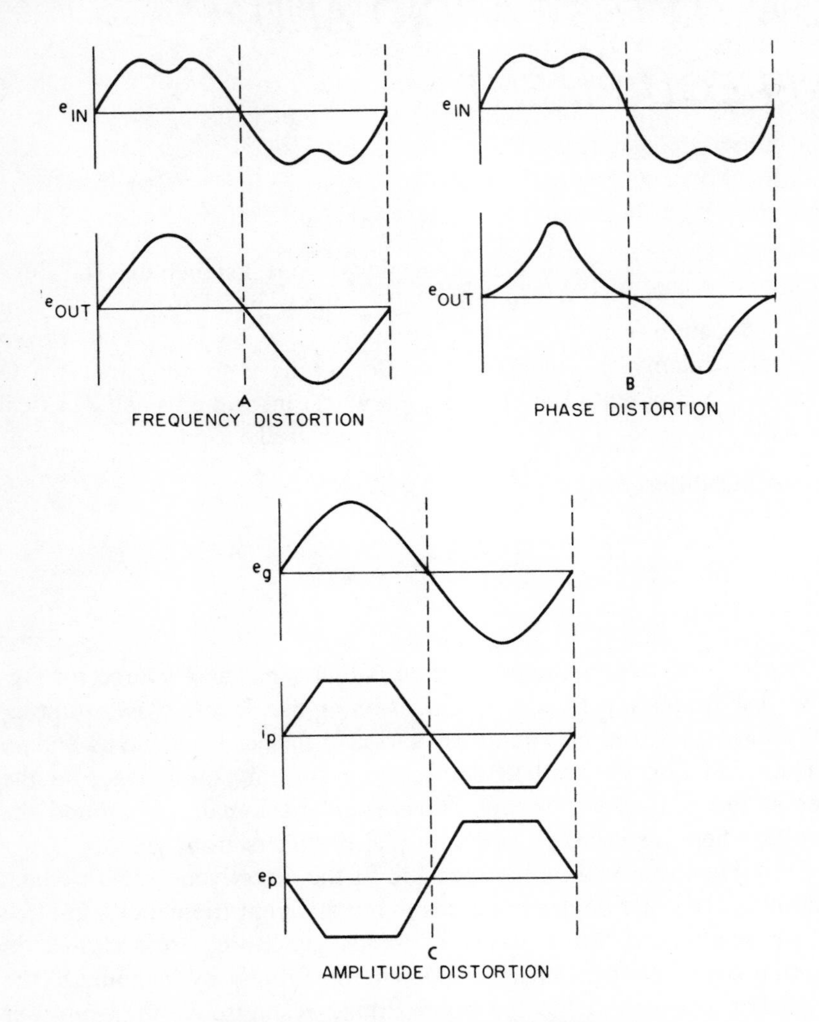

Figure 13.1 Types of Amplifier Distortion

Amplitude distortion is what happens when the amplitude of the output signal is not a linear function of the input amplitude. This means that the output waveform does not have the same form as the input wave. For example, when a strong sine-wave signal is clipped as shown in Figure 13.1 (a) because the operating point of one of the amplifier stages has changed, we have amplitude distortion.

Frequency distortion occurs when the amplifier fails to give equal amplification to all audio frequencies. In Figure 13.1 (a) the input signal to an amplifier contains the fundamental and third harmonic, but only the fundamental appears in the output.

Phase distortion is where the phase of the output signal is displaced relative to the input. If all frequencies were displaced by the same phase angle, it wouldn't matter, but they are not. Consequently, the phase angles between sounds of different frequency are altered as they pass through the amplifier, resulting in alteration of the quality of the sound. This is illustrated in Figure 13.1 (b), where the third harmonic has been shifted more than the fundamental, completely altering the waveform.

Harmonic distortion is produced when harmonics of fundamental tones are generated as the audio signal passes through the amplifier system. Harmonics are integral multiples of the fundamental frequency. For example, harmonics of 60 hertz are 120 hertz, 180 hertz, 240 hertz, etc., and are called the first, second, third (and so on) harmonics. Harmonics are weaker than fundamentals, of course, but they produce severe distortion as shown in Figure 13.2. Harmonic distortion always accompanies amplitude distortion.

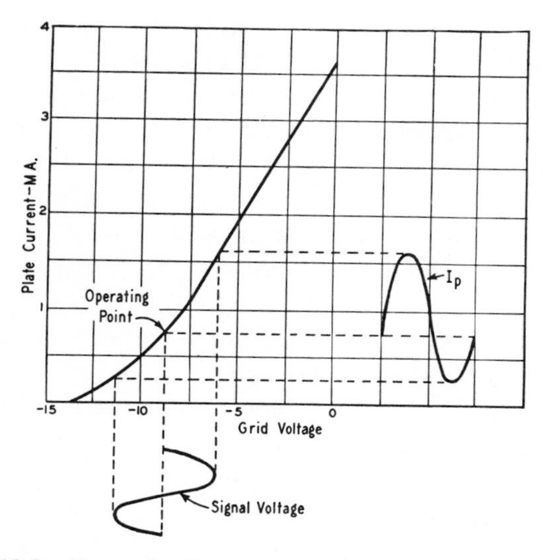

Figure 13.2 Harmonic distortion resulting from choice of an operating point on the curved part of the tube characteristic. The lower half-cycle of plate current does not have the same shape as the upper half-cycle.

Intermodulation (IM) distortion, from interaction (heterodyning) between two different frequencies, produces spurious signals with the sum and difference frequencies of the original signals. For instance, if IM distortion is produced between a 1000-hertz and a 100-hertz signal the spurious frequencies generated will be 900 hertz and 1100 hertz. In severe cases, a whole series of spurious beats are produced, as many different frequencies, including harmonics, react with each other.

Since the tones of speech and music are composed of fundamentals and harmonics anyway (Figure 13.3), you can see how garbled they can become in a poor amplifier as the original tones are distorted in these various ways. Tones with many harmonics, such as the sharp sounds of percussion or plucked strings, have waveforms with steeply rising fronts that may also shock some circuits into oscillation. The oscillation is usually a damped oscillation that adds to the signal fed into the amplifier, and in some cases will cause a very definite ringing sound in the output. Transient oscillations like these that die away in a few cycles may be caused by feedback loops acting as RC oscillators, or by self-resonant transformer or loudspeaker windings.

Continuous oscillation at a frequency above the range of hearing may be caused by a defective feedback network or decoupling capacitor. Although inaudible themselves, these oscillations may, nevertheless, overload the amplifier, causing amplitude distortion in the desired signal. In some cases they may cause IM distortion by beating with the desired signal, to give spurious audible tones in the output. This type of oscillation may break out only with a strong signal, because there is insufficient positive feedback from a weak signal to excite it. It is visible on the oscilloscope as a modulation envelope on the audio waveform.

Hum is another spurious signal that may be present in a hi-fi system. It often gets in from the turntable via an ungrounded shield, reversed connections, etc.

The *noise level* in the amplifier may also rise to audible level with aging tubes, overheated components, or demodulator defects in FM receivers.

Wow is caused by uneven speed of the turntable, resulting in untrue tones due to erratic rotation of the record.

Rumble is a low-frequency vibration superimposed on the signal at the pickup, caused by the motor driving the turntable.

Record scratch can result from a worn stylus, improper tone-arm pressure, improper mounting of the cartridge, dust on the record, scratches and so on.

It is obvious from the foregoing that hi-fi systems may develop quite a number of defects that degrade performance so that it's no longer hi-fi.

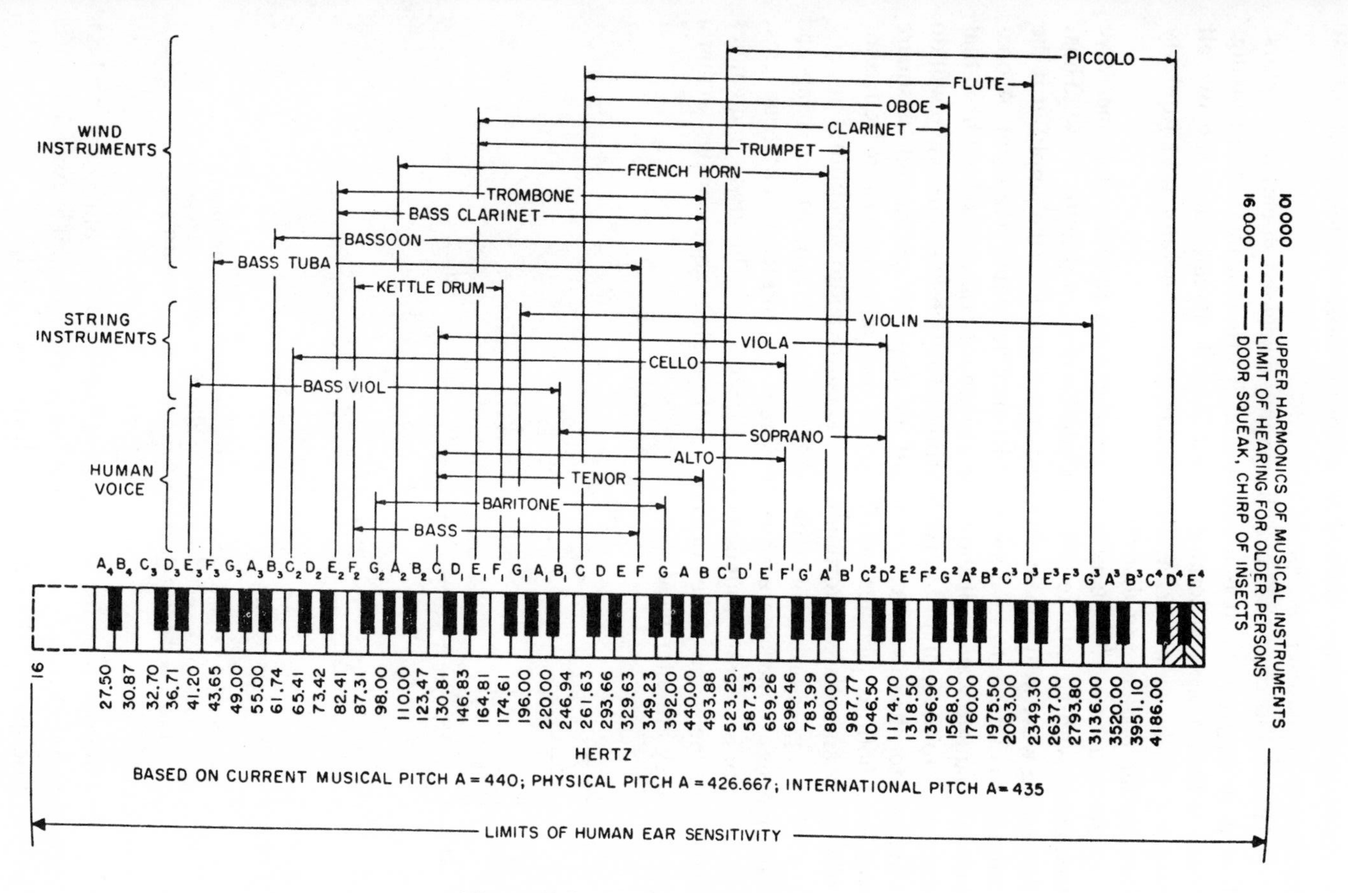

Figure 13.3 The Sounds of Music

In troubleshooting these problems, one of the most valuable pieces of test equipment is the oscilloscope.

The oscilloscope has been thoroughly discussed in earlier chapters. Just as in other areas, it is a very useful instrument in audio servicing. Used with a sweep generator or test record, it can show the overall response of the amplifier. With a square-wave generator it can quickly identify the presence of distortion.

As you know, a square wave consists of a fundamental sine-wave frequency with an infinite series of odd harmonics (Figure 13.4). Thus, the square wave will appear distorted at the output of the amplifier if the amplifier is unable to give all of its frequencies equal treatment. A one-kilohertz square wave will show how the amplifier handles them from that frequency to the limit of human hearing; a fifty-hertz square wave will do the same for the lower frequencies. If the amplifier can reproduce square waves of both frequencies satisfactorily, you can consider it to be free of frequency and phase distortion over the entire audio spectrum.

The higher-order harmonics determine the shape of the rising and falling edges of the square wave. If there is loss of the higher frequencies, those portions become rounded, as in Figure 13.4. The lower frequencies determine the shape of the horizontal portion of the square wave, so that it sags if they are attenuated.

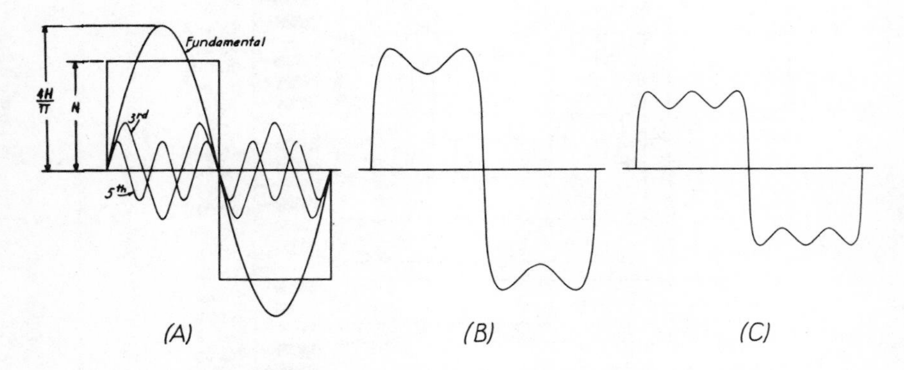

Figure 13.4 Composition of a square wave, showing the first three harmonics, and the approximation to a square wave as the first two and the first three odd-harmonics are combined.

The output of your square-wave generator should never be high enough to overdrive any of the amplifier's stages, and it is essential to use an oscilloscope that does not itself distort the signal.

The only items of test equipment used in audio and hi-fi testing that have not already been described in earlier chapters (or are adaptations of such equipment) are distortion and wave analyzers, and VU meters.

Components present in the output of an amplifier that were not present in the input are distortion. A distortion analyzer operates by suppressing the fundamental input frequency and measuring the residual distortion elements.

An audio oscillator generating a pure sine wave is connected to the input of the amplifier, and the distortion analyzer is connected to its output. The distortion analyzer (see Figure 13.5) is an AC voltmeter with the addition of a tunable rejection amplifier that may be switched in or out as required. (The impedance converter matches the low output impedance of the amplifier to the high input impedance (one megohm) of the distortion analyzer.)

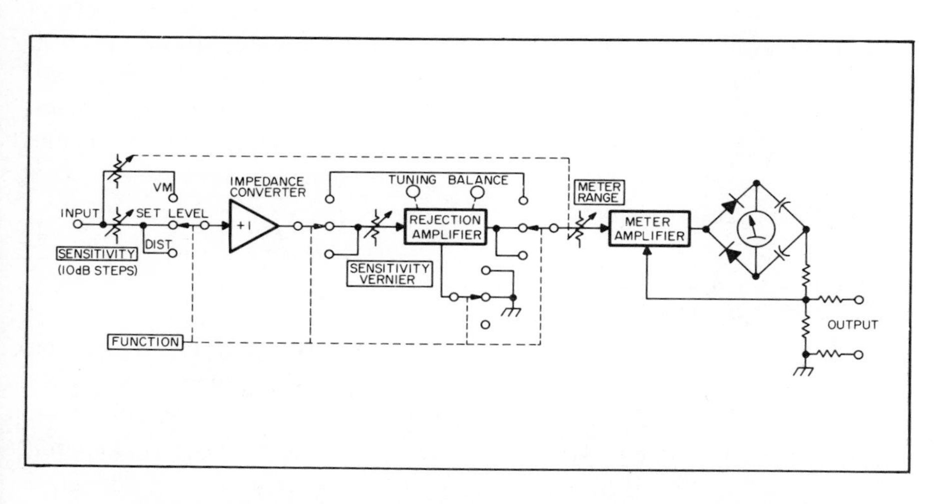

Figure 13.5 The distortion analyzer functions as a broadband calibrated ac voltmeter in "voltmeter" mode and as a signal level indicator in "set level" mode. In "distortion" mode, rejection amplifier can be tuned to suppress fundamental frequency of input signal, permitting comparison of distortion component's level to total signal level. *(Courtesy Hewlett-Packard Co.)*

You begin by placing the FUNCTION and METER RANGE switches in the SET LEVEL position, which bypasses the rejection amplifier, and then you adjust the SENSITIVITY control to set the meter pointer at the reference level (100 percent or 0 dB), which is the level of the total signal.

Then you reset the FUNCTION switch to DISTORTION, and adjust the TUNING and BALANCE controls for a null (lowest possible reading on the meter), switching the METER RANGE control downwards for greater accuracy. This tunes out the fundamental (input) frequency, so that the meter now reads the percentage or decibel level of the distortion elements that remain.

The meter is calibrated to read the RMS value of a sinewave, although responding to its average value. Usually, this is close enough for practical purposes, but if you need to know the *true* RMS value of the residual waveform, you can get it by connecting a true-RMS voltmeter to the OUTPUT terminals.

A wave analyzer is seldom used in audio work, but it is similar to the distortion analyzer, except that it has a narrow bandpass filter instead of the rejection amplifier, with which it rejects all frequencies other than the selected one.

In this way, it is tuned to each individual component of a complex or distorted waveform to measure its amplitude. It can measure the total harmonic distortion by obtaining their root-mean-square sum, expressed as a percentage of the fundamental frequency:

$$\text{Percentage Distortion} = \frac{100\sqrt{H_1^2 + H_2^2 + H_3^2 \ . \ . \ . \text{etc.}}}{F}$$

where F is the amplitude of the fundamental frequency and H_1, H_2, etc., the amplitudes of the harmonics.

Wave analyzers may be of the tunable selective-circuit type or of the heterodyne type. In the first-named (Figure 13.6a), the tunable amplifier is first tuned to the fundamental frequency, and the attentuator is then adjusted for a fullscale or 100-percent reading on the VTVM. The tunable amplifier is then tuned to the frequency of each harmonic component to be determined. This is then read directly as a fraction or percentage of the fundamental amplitude.

Figure 13.6a Wave Analyzer with the Timable Selective Circuit

The heterodyne type of wave analyzer (Figure 13.6b) resembles an AM radio, with a mixer stage and a very highly selective multistage amplifier resembling an IF amplifier. The frequency of the local oscillator is adjusted so that the heterodyne frequency produced by mixing its output with the desired component of the wave being analyzed is the same as the resonant frequency of the selective amplifier. Thus, the frequency of each component being analyzed is transformed to the same "IF" for measurement. Just as in the case of the superheterodyne receiver, this wave analyzer has excellent resolution and sensitivity.

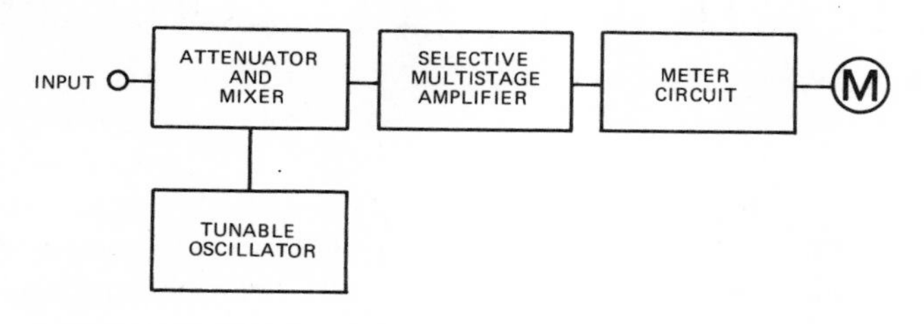

Figure 13.6b Wave Analyzer with Heterodyne Circuit

VU Meter

The sounds of speech and music are very complex, so that in their corresponding electrical signals there are rapidly fluctuating amplitudes. This creates special problems in measuring power output. Apparent power (in watts) is RMS voltage multiplied by RMS current but, as we saw in Chapter 2, a meter will read true RMS values only under certain conditions. Consequently, a special unit called a *volume unit* was invented for these measurements. A volume unit is the same as one decibel above or below the zero level, which is one milliwatt dissipated in a resistance of 600 ohms. A meter calibrated to read in volume units is called a VU meter. The reading on a VU meter represents the average of the instantaneous power of the audio signal, but it only coincides with the RMS power level if its waveform is a constant-amplitude sinusoid.

Audio power can also be measured as peak power, which is the maximum instantaneous power of the signal, equal to peak voltage multiplied by peak current. The meter circuit must be modified to provide capacitor storage of the peak values of voltage and current, as in the peak-responding voltmeter described in Chapter 6.

14

CALIBRATION OF TEST EQUIPMENT

In Chapter 1, we discussed the principles of calibration of test equipment, and explained that we can only rely on equipment that is properly maintained, and of which the limits of possible error are known.

This is done by evaluating the performance of the instrument being calibrated against that of a standard instrument whose accuracy is greater. For example, when calibrating a DC voltmeter you would use a DC voltage standard that gives precise selected output voltages, from which you can tell how accurately the voltmeter reads them. Similar procedures are used for instruments that measure AC voltage, resistance, frequency, and so on.

Suppose you want to calibrate a VTVM which has an accuracy of ±3 percent full scale for both DC and AC voltage, and ±10 percent of midscale for resistance. The best way of doing this is by using a *voltmeter calibrator*, such as the Hewlett-Packard Model 738BR (Figure 14.1 (a)), which provides both DC and AC voltages with an accuracy at least ten times better than the VTVM; and a decade resistor with an accuracy not less than ±1 percent. You should also have a variable *autotransformer*, such as a "Variac," to adjust the line voltage to the nominal 115 VAC and also to check that the VTVM operates satisfactorily with low and high line voltages.

You connect the VTVM to the calibration test equipment as shown in Figure 14.1, but before applying power you must adjust the meter pointer precisely to zero, using the mechanical-zero screw on the meter panel. Then turn on the power and allow about 15 minutes for warm-up. (In a calibration laboratory, the 738BR would be on all the time, of course.)

Set the function selector switch on the VTVM to +DC volts, and connect the DC probe and common clip together. Adjust the zero-adjust control so the meter indicates zero when the selector switch is in either the plus or minus position. This adjustment should suffice for all DC voltage ranges. However, if you should get an out-of-tolerance reading during the following procedure, you should recheck the zero for that range and readjust if necessary. This may save you from making unnecessary internal adjustments.

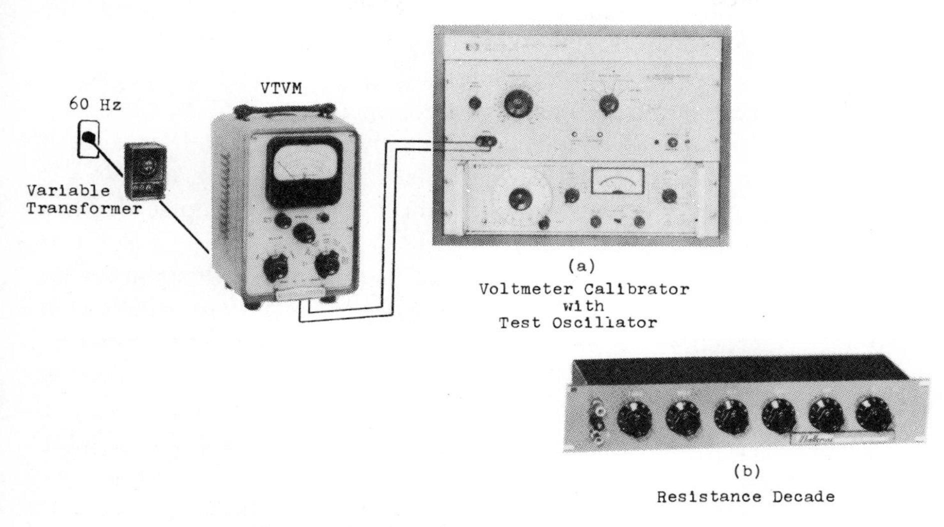

Figure 14.1 How to Calibrate a VTVM

Now you are ready to calibrate the DC voltage ranges. Connect the DC probe and common clip to the output of the voltage calibrator, and set the VTVM range switch to each range in turn. At each position set the voltage calibrator for an output equal to the full-scale value for that range. For instance, with a VTVM with the ranges shown below you would set the voltmeter calibrator accordingly, and would expect the voltmeter to read between the limits shown. These limits are for a ±3-percent meter; for other meters they will be as applicable.

RANGE SWITCH	VOLTMETER CALIBRATOR	ACCEPTABLE LIMITS MIN	(Volts)	MAX
1V*	1V	0.97		1.03
3V*	3V	2.91		3.09
10V	10V	9.7		10.3
30V	30V	29.1		30.9
100V	100V	97		103
300V	300V	291		309

*On this range, after performing check with positive voltage, reverse polarity of applied voltage and reset range switch to negative.

If the VTVM has a 1000-volt range, you'll need a high-voltage supply to check it as the voltmeter calibrator does not go that high. An indication between 970 and 1030 volts is satisfactory. Then reconnect VTVM to the voltmeter calibrator.

Write down each reading as you make it until you have done all the ranges. If any of your readings are outside the acceptable limits, and rechecking the zero for that range doesn't help, you should then adjust the internal calibration potentiometer. Some compromises will be necessary to get all readings within tolerance, but usually this will do it. However, if it is too far out, it must be because the range resistor in the VTVM attenuator for that range has changed value, and must be replaced. Then the calibration must be done again.

Your next test is for the linearity of the meter. For this you set the VTVM range switch to 1 volt, and then adjust the voltmeter calibrator for the following outputs. Each reading must be within the limits shown.

VOLTMETER CALIBRATOR OUTPUT	ACCEPTABLE LIMITS OF VTVM READING (Volts) Min.	Max.
0.8	0.77	0.83
0.6	0.57	0.63
0.4	0.37	0.43
0.2	0.17	0.23

Note that the limits in each case are the nominal values ± 0.03 volt (3 percent of 1.00 volt), since the tolerance is 3 percent of the full-scale value of the range, not of the reading.

If a reading is outside these limits, it indicates a defect in the meter movement, which must therefore be repaired or replaced.

Calibration of the AC ranges is done in the same way as for DC except that the voltmeter calibrator is adjusted for an AC output at 400 hertz, and the AC ZERO ADJ control is adjusted with the AC probe and common clip shorted together. You then connect the AC probe to the voltmeter-calibrator output terminals and check the ranges as for DC. Don't forget to read the values on the AC scales, if different from the DC ones (they are often colored red).

The voltmeter calibrator obtains the AC calibration signal from the test oscillator mounted in the same cabinet (Figure 14.1 (a)). When you have finished calibrating the AC ranges at 400 hertz, you should then check the frequency response at other frequencies. This is done by adjusting the 400-hertz output of the voltmeter calibrator to give a reading of exactly 0.9 volt on the VTVM, and then, maintaining the signal amplitude at the same level, adjust the test oscillator to the following frequencies:

60 hertz
100 hertz
1,000 hertz
10,000 hertz
100,000 hertz

At each frequency the VTVM must indicate between 0.87 and 0.93 volt (±3 percent of full-scale, of course).

While the VTVM is still connected for 100 kilohertz, adjust the Variac for a line voltage of 103 volts. In each case the VTVM reading must be between 0.87 and 0.93 volt. This checks the ability of the VTVM to cope with power-line variations.

RESISTANCE CALIBRATION

Disconnect the VTVM from the voltmeter calibrator, and set the SELECTOR or MODE switch to OHMS, and the RANGE switch to X1. Adjust the OHMS ADJ control so that the meter reads full scale.

Connect the ohms probe to the common clip. Adjust the ZERO ADJ control for an indication of zero on the ohms scale. Then go back and readjust the OHMS ADJ, if necessary. You may have to readjust both controls slightly, if they interact.

Connect the ohms probe and common clip to the terminals of a decade resistor (Figure 14.1(b)). Set the RANGE switch to the positions listed in the following table. At each position adjust the decade resistor for the values shown. Your VTVM should then read within the limits given (± 10 percent of the nominal value):

RANGE SWITCH	DECADE RESISTOR	ACCEPTABLE LIMITS (Ohms) Min	ACCEPTABLE LIMITS (Ohms) Max
X1	10 Ω	9	11
X10	100 Ω	90	110
X100	1,000 Ω	900	1,100
X1K	10,000 Ω	9,000	11,000
X10K	100,000 Ω	90,000	110,000
X100K	1 MΩ	900,000	1,100,000

If you have a X1M range, and your decade resistor box doesn't go any higher, substitute a 10-megohm resistor with an accuracy of not less than one percent. The VTVM reading should be between 9 and 11 megohms.

The frequency with which calibration is done depends upon the stability of the equipment. This is ascertained by experience. The purpose is to maintain the performance of the equipment within the limits of accuracy specified for it. Therefore, the interval between calibrations will depend upon how fast equipment parameters drift from their calibrated values. The manufacturer often supplies this information to the individual user. Large industrial firms usually determine their own, which may vary according to the use of the instrument.

In the foregoing example, we went through the procedure for calibrating a three-percent voltmeter, using an instrument at least ten times as accurate. If we were calibrating a more accurate voltmeter, such as a DMM, we should use a voltage standard with a proportionally greater accuracy. If we could not get one with ten times the accuracy of the instrument we were calibrating, we'd use the most accurate available, but we'd have to "shade" the tolerance limits as explained in Chapter 1.

One of the most precise DC voltages standards used for calibration of more accurate voltmeters is the Model 332 (A or B) manufactured by the John Fluke Mfg. Co., Inc. This voltage calibrator provides a DC output of 0 to 1111.111 volts at 0 to 50 milliamperes, with an accuracy of ±0.003 percent. The reference for the output voltage is a specially-selected zener diode with a constant-current source, both of which are enclosed in a temperature-controlled oven. The power switch on this instrument has a standby position, in which power is supplied to this circuit when the rest of it is inactive. In this way the instrument is kept ready for use at all times, and warmup drift is avoided. The actual drift under laboratory conditions will not exceed 0.0025 percent of setting ±10 microvolts per 6 months.

In addition to the positive and negative output terminals on this voltage calibrator, there are also positive and negative SENSE terminals. The latter are normally connected by metal links to the corresponding output terminals. However, if the load connected to the output terminals draws more than negligible current, the potential drop along the leads connecting it to the output terminals cannot be ignored if the high accuracy of which the instrument is capable is to be maintained. For instance, suppose you have a load that draws 50 milliamperes connected with two AWG #30 leads, each 3 feet in length. This will result in a voltage drop of 32 millivolts at 1000 volts, which is several times the maximum error allowed. To avoid this, you remove the links connecting the SENSE and OUTPUT terminals and connect additional leads from the load to the

SENSE terminals. Since practically no current flows in the SENSE leads, the voltage *across the load only* is measured, and the voltage drop in the OUTPUT terminal leads has no effect.

The most accurate voltage references found in laboratories are standard cells. There are two types: saturated and unsaturated. The saturated is the more accurate. Given at least three weeks to stabilize at a temperature that does not vary by more than 0.02 degree Celsius, under laboratory conditions, and provided it is not jarred, tilted or otherwise mishandled, its voltage will never vary by more than five parts per million from the measured value. This value, which varies slightly with each cell, is nominally 1.018 volts.

An unsaturated standard cell is rather more rugged, but with lower accuracy. It requires only a few hours to stabilize after being moved, and tolerates temperatures ranging between 22 and 25 degrees Celsius. It must also be handled with care, and its accuracy is ±0.002 percent. The nominal value is 1.019 volts.

Drawing any current at all from a standard cell will destroy it. Even measuring its voltage may be fatal (however, this is always stated on the label). So what use is it? It is used for comparison with some other source to ascertain the other source's exact voltage. For example, in the potentiometer setup shown in Figure 14.2 it is used to calibrate the working battery, so that the bridge-voltage measurements made with it have the accuracy of the standard cell and the resistances, without actually using the standard cell.

AC voltage calibrators with high accuracy are also required for the AC ranges of DMMs and other more accurate AC meters. Usually, this instrument is called an audio-voltage calibrator because its frequency range covers the band from 10 hertz to 100 kilohertz or thereabouts. The block diagram in Figure 14.3 illustrates the Hewlett-Packard Model 745A, which also employs an ultra-stable zener diode in a temperature-controlled oven. This is the reference for two voltages: +9.9 volts and −9.9 volts, that between them form a square wave. The accuracy of the RMS value of this square wave is approximately ±0.001 percent.

A second oscillator generates a sine wave, tuned by the front-panel frequency control, which is amplified to 100 volts and applied to the output attenuator. A sample of this voltage is compared to the square-wave signal in a comparator circuit, from which a feedback signal goes back to control the output amplitude of the sine wave oscillator signal. Over the frequency range from 50 hertz to 20 kilohertz, the accuracy of the sine-wave output voltage is ±0.02 percent, and the drift does not exceed 0.01 percent over a six-month period.

Resistance calibration precision is improved by using standard resis-

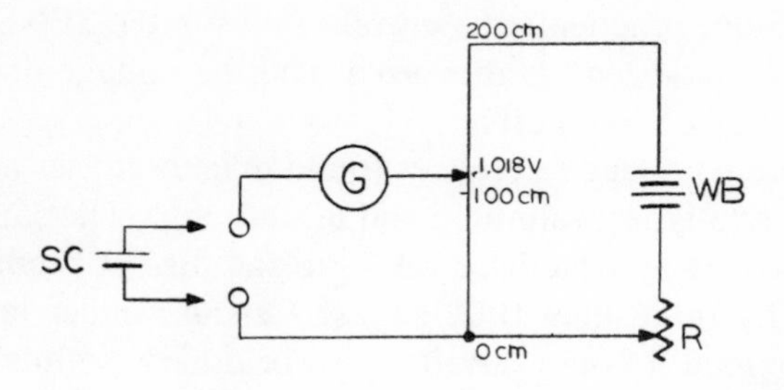

A resistance wire 200 cm long is connected across the 2-volt working battery WB. A sliding contact picks off voltages from 0 to 2 volts along the wire according to the length of wire between the position of the sliding contact and the lower end of the wire. For instance, at 100 cm from the bottom the potential will be 1.00 volt. The sliding contact is set at 101.8 cm, where the voltage will be 1.018 volts, which is the voltage of the standard cell SC. If this cell is now connected to the input terminals there should be no deflection of the galvanometer because the two voltages balance. If a slight deflection does occur, the rheostat R is adjusted until it is eliminated. This makes the potential exactly 2 volts between the 200 cm and 0 cm. Voltages along the slide wire will now correspond to the length in centimeters between the sliding contact and zero with the same accuracy as the standard cell, assuming uniform resistance along the slide wire.

An EMF to be measured is now connected to the input terminals in place of the standard cell, and the sliding contact is reset until the galvanometer again shows no deflection. The position of the sliding contact gives the value of the "unknown" EMF.

In a practical instrument a 200-cm slide wire would be inconveniently long, so part of it is made in the form of one or more dial resistors (shown below), and the slide-wire portion is made circular, sometimes with more than one turn, so that it takes up less space. Additional ranges can be provided by internal shunts, or by an external voltage divider. The latter permits the measurement of higher potentials, and is known as a "volt box."

Figure 14.2 Basic Potentiometer Circuit

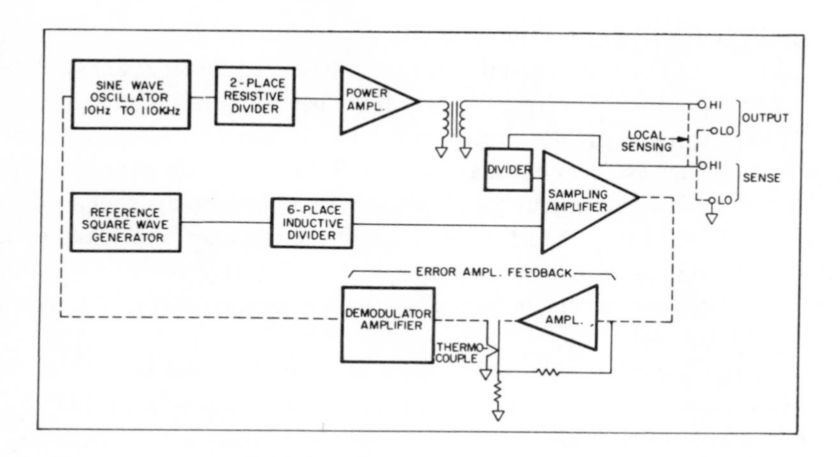

Figure 14.3 Audio-Voltage Calibrator

tors. The most accurate are fixed lengths of resistance wire immersed in oil in a cylindrical container, which allows for the insertion of a thermometer, since the exact value of the resistance depends upon the temperature. A resistor of this type may have an accuracy of one part per million, or 0.0001 percent of its value. These resistors are normally provided with four terminals so that voltage measurements can be made across them without having to worry about potential drop in the current leads.

Decade resistors with higher accuracy than the one used to calibrate the VTVM are available, and are usually used for calibrating DMMs and other instruments with similar accuracy.

As a general rule, all signal generators are calibrated for frequency with electronic counters, and with AC voltmeters for amplitude of the output. The purity of the waveform is verified with a distortion analyzer (described in Chapter 13).

The accuracy of the counter is very great. Indeed, counters are among the most accurate instruments in the laboratory. As we have seen already, an electronic counter is an instrument for comparing an unknown frequency or time interval with a known frequency or time interval. The precision with which it does this depends primarily upon the stability of the known frequency.

In most counters this is derived from its internal time-base oscillator, which is controlled by a quartz crystal maintained at a constant temperature in a crystal oven. The accuracy is expressed as ± 1 count ± time base accuracy. The best crystal-oven time bases have accuracies of the order of 5 parts in 10^{10} (1 part in 2 billion) per 24 hours after warmup.

However, even greater precision can be achieved by using an external time base such as a cesium-beam frequency standard with an accuracy of ± 1 part in 10^{11} (1 part in a hundred billion), or an atomic hydrogen maser accurate to 1 part in 10^{12} (1 part in a trillion). These are two primary standards that provide access to invariant natural frequencies in accorance with the relationship of quantum mechanics. They may be used as external frequency sources for counters.

The time interval of atomic time is the international second as defined in Chapter 1. This is not the same as the universal time scale (UT2) second, which is related to the earth's rotation. That second is slightly longer than the atomic second, so that transition from the former Universal Time Scale to the Atomic Time Scale results in some compromises. By international agreement the Bureau International de l'Heure in Paris determines the amount of offset between the two scales for the year, which is of the order of -300×10^{-10}, and each country then adjusts its own Standard Time accordingly. In the U.S. the time interval broadcast by National Bureau of Standards stations WWV and WWVH is an approximation of UT2, ("Coordinated Universal Time"), but WWVB (60 kilohertz) broadcasts the atomic second without offset. Calibration laboratories use these broadcasts to synchronize their own standards with the U.S. Frequency Standard, which has an accuracy of parts in 10^{12}, and is among the world's most accurate. Where the calibration laboratory is part of a large manufacturing plant, the output of its frequency standard is distributed throughout the plant, to be used as an external time base in high-performance counters, thereby upgrading their accuracy to that of the standard.

The calibration of oscilloscopes involves the time base (sweep), vertical and horizontal amplifiers (input attenuators), calibrator, and in some cases other additional functions.

It is usual to begin by checking the operation of the front-panel controls to verify the proper functioning of those that adjust the position, focus and intensity of the electron beam, and to continue by measuring the calibrator output at each setting to ensure it is within the tolerance specified. Where the calibrator signal is a square wave generated by a multivibrator, the most accurate way of measuring it is to disable the multivibrator (by removing the tube, for example). This stops the multivibrator output from switching between the upper and lower voltage levels of the square wave, so you can use a DC voltmeter to measure the

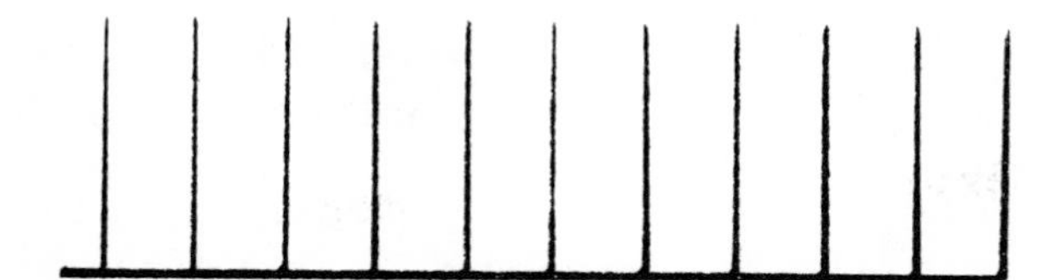

Figure 14.4 Typical Oscilloscope Display Produced by Time-Mark Generator. On the Time-Mark Generator you select the display corresponding to the oscilloscope time-base range you wish to calibrate. The time marks (vertical teeth) should fall exactly on the vertical graticule lines. Adjustments in the oscilloscope are provided for correcting the sweep speed if they do not.

upper level, which is much more accurate than measuring the square wave with an AC voltmeter.

After making any necessary adjustment, the calibrator signal can then be used to check the performance of the vertical and horizontal amplifiers. The gain of these should be such that the amplitude of the waveform on the screen is the product of the calibrator signal and the input attenuator setting, expressed in centimeters. If the gain is incorrect, it can be adjusted by means of the internal gain control adjustment.

To check the triggering you apply a low-amplitude sine wave at about one kilohertz to the vertical input, adjusting it for a display of two millimeters, and check that a stable display is produced with both positive and negative internal triggering at the same setting of the triggering level control.

Checking the sweep time base accuracy is best done with a Time Mark Generator. This is connected to the vertical input of the oscilloscope, and produces a comb-shaped trace, as shown in Figure 14.4. The markers are supposed to be exactly behind each vertical graticule line, so if they are not, the time base may have to be adjusted. Each setting of the TIME/CM switch is checked in this way. Measurements are always made between the second and ninth markers to avoid any slight non-linearity at the ends of the trace.

Finally, the high-frequency response of the vertical amplifier is checked. The best way to do this is with a pulse generator, applying a fast-rising pulse to the oscilloscope vertical input. To preserve the characteristics of the pulse be sure to use a properly terminated coaxial cable. The way this is reproduced on the screen tells you much about the performance of the amplifier, as we saw in Chapter. 9. There should be no excessive roll-off, overshoot or ringing on the front corner of the displayed waveform. Last of all, you measure the risetime between the 10 and 90 percent points of the leading edge of the pulse, using the oscilloscope magnifier to get a broader waveform. Any degradation of the waveform must be due to the limitations of the vertical amplifier.

RECORDS

Not the least important of the functions of a calibration laboratory is the keeping of proper records. When an instrument is calibrated, its condition before and after should be documented so that its performance may be compared from calibration to calibration, and a history compiled that shows its general reliability and particulars of troubles requiring repair. In this way troubleshooting is made easier, since many instruments have an inherent tendency to go in for repeated failures of certain components. Such a history is also a guide to the desirability, or otherwise, of replacing the instrument with a successor of the same type or manufacture when the time comes.

Such a history, by telling you how much the instrument's performance parameters drift between calibrations, enables you to judge how long an interval to allow. It can also be filed in such a way as to notify the user when calibration is due. This is done usually by calendar date when the instrument is in constant use, but may be done on operating time if used intermittently. In the second case, an elapsed time meter that runs only when power is applied to the equipment will indicate either how much running time has been accumulated or how much is left.

Most large companies now use a computer to keep these records, and the computer automatically adjusts the calibration intervals and issues recall notices in accordance with the data it receives. This has proved to be a very efficient method, since flexible calibration intervals take full advantage of the capability of the instrument, minimize down time and eliminate unnecessary work, including saving wear and tear to the equipment.

15

HOW TO SAVE TIME WHEN TROUBLESHOOTING

A troubleshooter is a person who locates and corrects a malfunction, which is defined as the failure of a piece of electronic equipment to function as it should. As a general rule, locating the malfunction is harder than correcting it. This chapter outlines a practical procedure for identifying the cause of a trouble by careful investigation of its symptoms. It is a logical process similar to that used by a doctor in diagnosing a disease.

It may also be compared to the method used by a detective in solving a crime, at least according to most murder mysteries. The clues he discovers lead him to suspect several people, and by a process of elimination he narrows them down to only one, whom he arrests. The troubleshooter also starts with a set of clues from which he determines the major functional unit or section of the equipment involved, and then performs tests to narrow the area of search to the circuit concerned, and finally isolates the individual component responsible.

Of course, he may not need to carry out the entire procedure. In simple pieces of equipment there is not so much to check. Sometimes the cause of the malfunction is glaringly obvious—the criminal is caught red-handed, so to speak. However, the following table represents the principles of troubleshooting, or malfunction isolation if you prefer:

STEP	PROCEDURE	TEST EQUIPMENT
A. DETERMINE SYMPTOMS	Determine precisely the way in which the behavior of the equipment departs from normal.	—
B. DETERMINE SECTION	Determine which major functional section of the equipment is involved from the evidence obtained in step A.	—

C. TEST CIRCUIT	By signal tracing in Section locate faulty circuit.	Signal generator, oscilloscope or voltmeter
D. TEST COMPONENT	Identify cause of problem by testing individual components in circuit.	Voltmeter, ohmmeter, tube/transistor tester, eyes, ears, nose, touch

A. Determine Symptoms

In order to determine precisely the way in which the behavior of the equipment departs from normal, you must know the normal operating characteristics of the gear, and be able to manipulate the operating controls as well as, or better than, the equipment operator himself.

Start by observing whether the operating controls were adjusted incorrectly, and set them all to normal or safe operating positions. If the trouble symptom is the result of a control-setting error, or control malfunction, you can remedy it by correcting the adjustment or replacing the control. If the controls are all right, their resultant effect upon the symptom may aid in determing the location of the trouble.

Symptom recognition occurs when you distinguish between normal operation and undesirable changes in equipment performance. A failure is recognized when there is a lack of performance information that would normally be seen or heard by the operator. Degraded performance is when the present performance deviates from the normal performance.

Symptom evaluation is the process of obtaining more information about a trouble symptom. You should record all data during this stage of troubleshooting because it permits a complete analysis of all pieces of information in relation to each other, enables you to assign to them their proper importance, and makes it possible to recreate specific performance characteristics when necessary.

You should pay special attention to the displays that the equipment itself produces, and compare these displays with your knowledge of how the equipment should perform. We saw in the chapter on television test equipment how important this can be in indicating the functional section of the set that is at fault.

B. Determine Section

A section is a group of circuits in which some specific electronic function is performed; hence it can also be called a functional unit. In major pieces of equipment there will usually be several sections, but in small ones there may well be only one, in which case you would omit this step.

The first step in deciding which section is indicated is to study the block diagram. The block diagram shows each unit as a rectangle, but does not tell you how its function is performed. Signal flow paths between units are represented by single lines, but no details are given about the actual connections. However, waveforms are sometimes given, in conventional or idealized form, without any values. In other words, the block diagram provides a general picture of the major functional units of the equipment and their important signal relationships, as in the example in Figure 15.1.

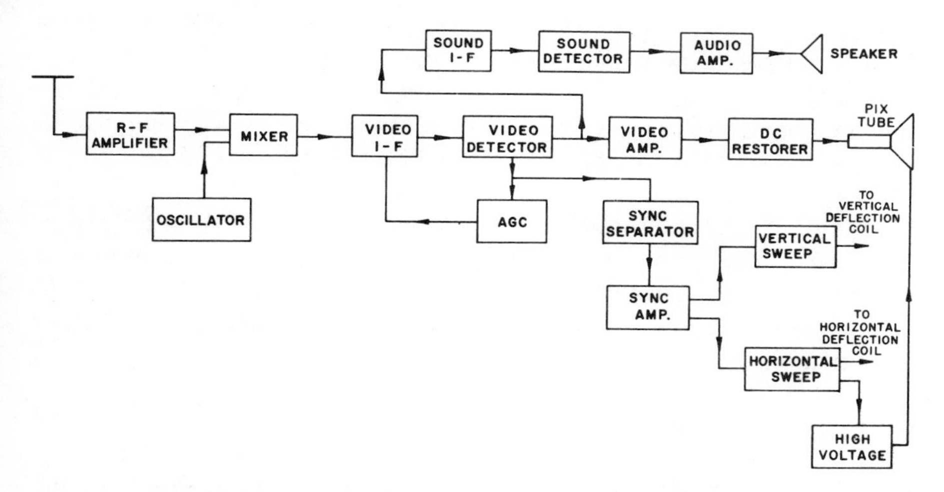

Figure 15.1 Block Diagram of a Typical Monochrome TV Receiver

The first step is to consider what units, as shown in the block diagram, could logically contain the malfunction. This is where your knowledge of the functioning of the equipment is extremely important. Equally important is the accurate determination of the exact nature of the symptoms. Without these two you cannot draw a logical conclusion at all.

In the block diagram in Figure 15.1, which section would you suspect if the symptom were, say, a single horizontal line across the picture-tube screen, while the sound was coming through normally? The audio output is good, therefore you can eliminate the sound section right away. The horizontal line all the way across the screen implies that you have full horizontal deflection but no vertical deflection. The horizontal deflection circuits are, therefore, free from suspicion. You cannot tell whether the picture is satisfactory in other respects as you cannot see it, but it seems reasonable to suppose that the power supply is working. You *do* know there must be something wrong with the vertical deflection section, because its output (picture height) is missing completely.

You can, therefore, mentally place a bracket to the right of the vertical deflection block, since you are sure the output is bad. Now you should ask yourself: If all the other sections from the antenna to the vertical deflection section were working properly, would I still have this symptom? If so, it is safe to assume for the moment that the tuner, IF, video detector and sync circuits are performing properly. Place a second bracket to the left of the deflection block. This, then, is where the problem is most likely to be.

However, if your symptom had been raster and sound but no picture you could not know without further tests whether the trouble lay in the tuner, the IF amplifier or the video detector and amplifier, since the picture signal has to pass through all of these, and the symptom of no picture, without any other evidence, could be caused by trouble in any of these.

This is why it is important to determine accurately what the symptoms are, since supporting evidence such as the presence or absence of sync signals (see Chapter 12) would be able to eliminate some of these sections—those in which the sync information accompanies the picture information.

C. Testing to Locate the Faulty Circuit

This step determines which circuit in the equipment contains the components that must be repaired to restore normal operation. The servicing block diagram, which is more detailed than the functional block diagram of Figure 15.1, provides a general representation of the circuit groups, and individual circuits as well, and their signal-path relationships, waveforms and test points. If there is no servicing block diagram available, the overall schematic for the equipment is used instead.

Tests are made using signal tracing or signal substitution. In signal tracing, the input signal is applied at a fixed point and the oscilloscope test probe is moved from place to place. (The test probe need not be that of an oscilloscope, of course, if some other piece of test equipment, such as a VTVM, would be more appropriate.)

In signal substitution, the opposite procedure is followed. The test probe is fixed and the input signal is applied from place to place.

Unless the symptoms already point to a particular circuit, it is usually quickest to go to a test point halfway between the good-signal input and the unsatisfactory-signal output, and see what you get. If the signal there is good, the trouble must be in the second half of the section; otherwise it has to be in the first half. In this way you halve the area of search.

In the vertical deflection section in Figure 15.2, the halfway point lies between the blocking oscillator circuit, in which Q1 is the active element, and the output circuit, where Q2 is the driver and Q3 the power output transistor. If the oscilloscope probe is applied to the base of Q2, a signal resembling Figure 15.3 should appear on the CRT screen. If it does not, the trouble must be to the left of this point. Conversely, if it does, the problem must be further on, to the right.

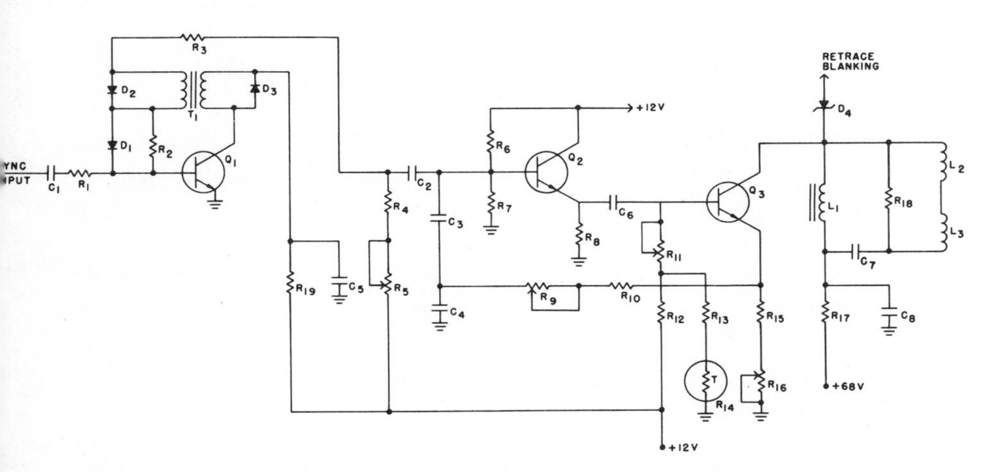

Figure 15.2 Vertical Deflection Circuit of Solid-State TV Receiver

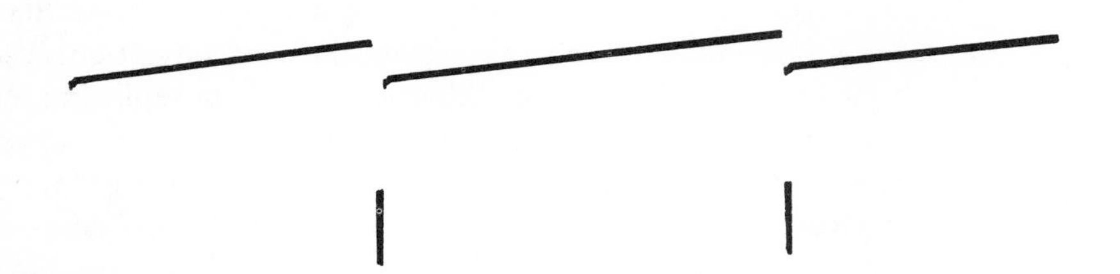

Figure 15.3 Signal on Base of Q2

Assuming the second indication to be the case, you subdivide the circuit to the right by transferring the oscilloscope probe to the base of Q3. Again, the presence of the proper signal would direct your attention to the right; its absence would point you to the left.

If the signal on the base of Q3 is absent, you can see why no vertical deflection is taking place. There is no input signal to Q3 to make it produce the output signal that drives the vertical deflection coils L2 and L3. Your next step is to identify the component that is responsible for the loss of signal. This step has enabled you to identify the circuit in which it must be located.

D. Testing to Locate the Faulty Component

You can eliminate Q2 quickly by checking the output signal on the emitter. If satisfactory, the trouble has to be the coupling capacitor C6. R8 must be all right, otherwise Q2 could not function properly. Similarly, troubles in the resistive divider network R11, R12, R13, and R14 would not stop the signal reaching the base of Q3, though they would affect the transistor's operation.

It's pretty certain that C6 is open. To verify this, bridge C6 with a good capacitor of approximately the same value. The resultant appearance of a proper picture on the screen would be confirmation of the correctness of your diagnosis.

One more question needs to be asked. Would the defect in this component be the sole cause of the problem? In other words, would replacing this component effect a complete cure? The answer in this case is obviously yes, but sometimes it's not quite as easy.

For example, suppose the signal had been present on the base of Q3. Several things might have prevented its appearance in the output. Q3 might have been defective, L1, R15, R16 or R17 might have been open, C8 might have been shorted. If R17 was found to be open, it might be because C8 was shorted. Either could be responsible for a failure of Q3 to operate. In every case, then, it is necessary to consider carefully whether other components could be involved. If a resistor has burned out because a capacitor shorted to ground, replacing the resistor without replacing the capacitor would only result in another burned resistor.

It is not the intention in this chapter to discuss methods of making repairs once the trouble has been identified. That part of the procedure is not troubleshooting, and does not belong in this book. What we have done is to present troubleshooting as a four-step process, which, if employed with logic and a proper understanding of the normal performance characteristics of the faulty equipment, will be the most powerful item in the repairman's tool kit.

INDEX

INDEX

N

O

P

Q

R

S